The Indispensable Excess of the Aesthetic

The Indispensable Excess of the Aesthetic

Evolution of Sensibility in Nature

Katya Mandoki

LEXINGTON BOOKS
Lanham • Boulder • New York • London

Published by Lexington Books
An imprint of The Rowman & Littlefield Publishing Group, Inc.
4501 Forbes Boulevard, Suite 200, Lanham, Maryland 20706
www.rowman.com

Unit A, Whitacre Mews, 26-34 Stannary Street, London SE11 4AB

British Library Cataloguing in Publication Information Available

Library of Congress Cataloging-in-Publication Data

Mandoki, Katya, 1947-
The indispensable excess of the aesthetic : evolution of sensibility in nature / Katya Mandoki.
pages cm
Includes bibliographical references and index.
ISBN 978-1-4985-0306-8 (cloth : alk. paper) -- ISBN 978-1-4985-0307-5 (electronic) 1. Senses and sensation. 2. Aesthetics. I. Title.
BH301.S46M36 2015
111'.85--dc23

2015005997

™ The paper used in this publication meets the minimum requirements of American National Standard for Information Sciences Permanence of Paper for Printed Library Materials, ANSI/NISO Z39.48-1992.

Printed in the United States of America

לחיים

Contents

Acknowledgments

I wish to thank the Hebrew University of Jerusalem for inviting me as Visiting Researcher at the Faculty of Humanities and for access to the wealth of material at Givat Ram and Mount Scopus Libraries, which has nourished my work from my student years at the philosophy department to the present. I would also like to thank Cambridge University Library for permission to consult the Darwin Archive's originals and for its fabulous Darwin webpage. My special thanks to Sistema Nacional de Investigadores (SNI) and Consejo Nacional de Ciencia y Tecnología (Conacyt) for supporting this research as well as to the Universidad Autónoma Metropolitana for the part time invested on sections of this work. I thank also the International Association of Aesthetics, the International Society for Biosemiotic Studies, and my postgraduate students for opportunities to discuss sections of this text. My appreciation to Cornell Lab of Ornithology, to National Geographic, and to the many contributors to YouTube who teach us to discover our world and know our animal neighbors in all their intelligence, emotional quality, and magnificence.

My gratitude to the publishers of the following articles for granting permission to reprint them in whole or part:

1. Mandoki, Katya. "Zoo-aesthetics; a Natural Step after Darwin." In Timo Maran (ed.), *Semiotica: Journal of the International Association for Semiotic Studies / Revue de l'Association Internationale de Sémiotique* Special Issue in Zoosemiotics, Volume 2014, Issue 198. Berlin: Mouton de Gruyter, 2014, pp. 61–91.
2. Mandoki, Katya. "The Indispensable Excess of the Aesthetic." *Filozofski Vestnik—Acta Philosophica* 20, no. 2 (1999): 173–179.

3. Mandoki, Katya. “Material Excess and Aesthetic Transmutation.” In *Economies of Excess: Parallax*. John Armitage (ed.). Issue 18, Vol. 7, No. 1 (Jan–March 2001), pp. 64–75. London: Taylor & Francis.

Prologue

Up, wherever that may be, is where aesthetics has remained since Plato set Beauty at the top of the World of Ideas, next to Goodness and Truth. "Elevated style" is how Longinus distinguished the sublime over the ordinary. The word "higher" is repeated four times through three consecutive sentences; eight times in the page in Georg Wilhelm Friedrich Hegel's lectures on aesthetics ("beauty of art is higher than nature. . . . And the higher the spirit and its productions stand above nature and its phenomena, the higher too is the beauty of art above that of nature. Indeed . . . a useless notion that enters a man's head is higher than any product nature." (Hegel 1975, 2) Even a down-to-earth pragmatist as John Dewey is tempted to place art at the very top of the mountain of everyday experience, doing and undergoing. (Dewey 1980, 3)

And there it lingers: for much of philosophical discourse the aesthetic remains at lofty heights. By looking at the world from above, this philosophy of verticality predictably locates the spirit at the top and matter below, artistic beauty up and natural beauty down, the rational above the emotional, the mind over the body. However, this enhancement of the aesthetic to such elevated peaks belies its original sin: the growing evidence that the sense of beauty actually comes from underneath, from our pudendum. Judging beauty may be less work of the spirit than of the body, less a matter of culture than of nature, and less a fruit of virtue than of lust. Let us then begin from the origin of any possibility of aesthesis in the living, the cell, the plant, the animal, and the human. We must consequently explore it from below.

We ought to explore it also from aback. We can not point forwards and join the avant-garde artists who, in their utopian manifestos by the early twentieth century, imagined themselves passionately leading humanity into the glorious triumph of Art. Instead we will look into our past and read backwards the traces left during billions of years by the inexplicable tenacity to survive and multiply. In our bodies we carry a legacy that includes not only organs for metabolism, respiration, and locomotion but also and especially sensory organs and neuro-cerebral particularities that allow us to detect and interpret the world for remaining in it as long as possible and passing on this gift to our offspring. Therefore, in addition to our genetic heritage, we have inherited an aesthetic legacy that configures our sensibility for valuing our world, but that at the same time inevitably condemns us to ignore all other forms of possible perception. Our senses, emotions, and preferences attest to an evolving aesthetics through our body, where else.

For undertaking this work it has been necessary to shake off if not altogether abandon some traditional dichotomies (mind-body, spirit-matter, emotion-thought, aesthetic-logic, nature-culture) that have restrained our mind at least since the Pythagorean school. As analytical categories these are essential but when taken literally as reality itself they become a burden.

Another unavoidable step requires overcoming the canon that restricts Aesthetics to the study of Art and Beauty, categories that are certainly relevant, but reduce this rich field of research to function merely as a Geiger counter of beauty or as an artwork certification device. Aesthetics is thus risking being replaced by statistical surveys of taste, golden section detecting software, or the wavelet digital decomposition method for artworks.

The full extra-artistic spectrum remains to be explored, no less than six of the seven rainbow colors, namely the aesthetics in seduction and mating, in rituals and celebrations, in the joy of achievement and admiration of excellence, in the figuration of myths and the pleasure of playing, and especially in the miraculous opening up to the world of every single creature.

We will here proceed simultaneously through two perfectly complementary directions: biosemiotics and evolutionary research. Semiotics is an endeavor whose origins we can trace to Greek antiquity in studying body symptoms as signs of disease. During the Middle Ages and the Renaissance it was practiced as a "theo-semiotics" for understanding the world as a constellation of God's indexes. During the twentieth century, by the so-called linguistic turn, semiotics became totally obsessed with verbal signs to the point of forgetting their natural and corporeal sources.

Von Frisch's spectacular findings on communication among bees, Peirce's pragmaticist view on semiosis, and von Uexküll's work on animal perception impelled the great semiotician Thomas Sebeok to declare himself a "frustrated biologist" and return semiotics to biology (Sebeok 1963; von Frisch 1955; von Uexküll 1992). A century earlier, Darwin began a semiotic reading of evolution when he decoded morphology and behavior in plants, animals, and humans as indexes of their past and context with greater diligence and acuity in interpreting traces than Sherlock Holmes in interpreting marks left by the felon on the crime scene.

Yet semiotics and evolution, merging from and into pragmatism, are not the main subject of this book, as recurring as they are along these pages: they simply serve as our tools to lead us in exploring the very problematic field of aesthetics. Surprisingly, already in the mid-eighteenth century, at the very foundation of aesthetics as an autonomous discipline, Alexander Baumgarten attempted also to follow a semiotic approach to pioneer this field. We can read in the table of contents of his unfinished *Aesthetica* that the second part of the Aesthetica Theoretica is divided into Heuristica, Semiotica, and Methodologia, if in a different sense from modern semiotics. (Baumgarten 1750, index)

We will proceed as follows: in the first chapter we will introduce the basics of evolutionary thought and the concept of aesthesis along a spectrum of degrees in sensibility. Chapter 2 is dedicated to biosemiotics, our second frame of reference for exploring aesthesis entwined with semiosis as in a Möbius strip.

Unearthing the explanatory potential of evolutionary aesthetics, David H. Mandel (1967, 11) wrote: "We begin with the dance of the electron and end with the dance of imagination. We begin with "inert" matter, to which we are related, and end with the creative powers of specially endowed human beings, the artists. Evolution, the miracle of growing sensitivity and increasing consciousness, has led us so inevitably to art that in a certain view it may finally be through aesthetics that we interpret the meaning of biology."

Hence, we have to begin with the dancing electron or even earlier, with the prankish photon. We must start at the *Orbis primus*, or inanimate (if not inert), physical world, where we will observe how the aesthesis–semiosis coupling evolved as a process of perception and signification. Passing to *Orbis secundus*, we follow the emergence of sensibility in chapter 3 from the cellular organisms to even more complex creatures. The next step will be to explore another threshold, *Orbis tertius*, where we deal in chapter 4 with how sensibility is expressed and modified when a critical mass of individuals assembles from early forms of collective organization in various species to human societies by means of common agreements, thus diversifying as well as accumulating cultural practices. Finally, in chapter 5 we inquire on the intimate link between excess and the aesthetic across these various levels as in signaling surplus and lavish displays of expenditure. Throughout this journey, the enigma of beauty will wink at us with new, unexpected faces.

Chapter One

Aesthesis

Zan tenyotiliuh xochime, zan tenyotiliuh cuicame.
—Nezahualcoyotl

"Let us at least leave flowers, let us at least leave song," writes the Aztec poet Nezahualcoyotl. Nature as flower and culture as song are the two images that, as butterfly wings, join to rest upon the aesthetics of this great poet from Texcoco. Our body withers during the brief instant that it is alive, but the soul remains in song if only someone is there to listen and sing it, as we now listen to Nezahualcoyotl. When a flower blossoms and a song is sung, what flows through them is aesthesis.

Why are we attracted to flowers? Why are we moved by songs? Why do rhythms captivate us so intensely? What is there in a voice that stirs us, freezes or encourages us? Can other creatures appreciate beauty? Why are fundamental particles of nature so intimately connected to questions of symmetry? Or, perhaps, why are we so attentive to questions of symmetry?

Evolutionary psychology proposes that the answers to a variety of seemingly unrelated mysteries must be found in our distant past two and a half million years ago during the Pleistocene when our hominid ancestors' corporeal morphology evolved to its present configuration. Darwin's explanation—random changes in organisms favorable to their survival and reproduction are retained while the harmful are lost by limiting the reproduction of their carriers—reinforced by DNA and molecular biology breakthroughs, games theory, computer simulation models, and population genetics consolidated a paradigm that extended its elucidatory power from the field of Biology to the Humanities.

"How extremely stupid [of me] not to have thought of that!" exclaimed Thomas Huxley when he realized that the simple formula of "random variation and natural selection" sums up evolution's basic mechanism.[1] Such formula would be, however, only part of the explanation and Darwin was the first to acknowledge it. We must consider other types of selection, namely sexual selection, organic selection, genetic, epigenetic, symbiogenetic, and behavioral, group, social, and cultural forms of selection that are issues still under debate.

The reverberations caused by the controversy surrounding nineteenth-century social Darwinism and twentieth neo-Darwinism from Lorenz's ethology and Wilson's sociobiology are still in the air. The differences within evolutionary theory are also many and intense, but they have the undeniable merit of keeping the discussion of humanities at ground, empirical level. Aesthetics are not exempt from these debates.

Since aesthetics is an outcome of nature and not a divine celestial gift, we must reflect upon it from the perspective of corporeality and its evolution. Some results of this new evolutionary approach to aesthetics are emerging and demanding a revision of their assumptions. For reasons explained throughout the text, the position assumed here is inverse to the approach taken by sociobiology, but also to structuralist positions that reject an evolutionary focus on culture: we will examine cultural expressions by considering their umbilical cord to biology in retrospective and not projecting biology upon culture from a prospective position. We will proceed by using the so-called reverse engineering method that Darwin tacitly applied to understand the conditions that shaped organisms by deducing from the design of a body organ the contexts and needs within which it evolved. This method, together with hard work of observation and very detailed, often tedious, annotation resulted in the most important paradigm revolution of the natural sciences to this day.

Nature is lavish and generous as well as frugal and precarious. Evidence on its wastefulness is blunt in natural disasters, in the almost infinite variety of existing or extinct species and in the bountifulness of seeds discarded by trees for only a few exemplars' fertilization. At the same time, the testimony of its practicality is displayed by the efficient, almost minimalistic and precise design of organisms. Our corporeality attests to the world we inhabit and to its past.

THE AESTHETIC COMPASS

> Our whole cubic capacity is sensibly alive; and each morsel of it contributes its pulsations of feeling, dim or sharp, pleasant, painful, or dubious, to that sense of personality that every one of us unfailingly carries with him. It is surprising what little items give accent to these complexes of sensibility.
>
> (James 1884, 192)

Focusing aesthetics from the body means focusing it from its evolution as a biological phenomenon since "nothing in biology makes sense except in the light of evolution" according to Dobzhansky's famous phrase. (1963) This seemingly obvious fact does not appear to be so whenever evolutionary approaches to aesthetics rush to solve puzzles with two major obstacles in tow: The first stems from projecting concepts from biology upon the field of culture without any mediation. The second is the traditional misunderstanding of the term that curses this discipline from its origins: its confinement to art and beauty. As recorded in Baumgarten's founding book in the mid-eighteenth century, aesthetics is defined as *scientia cognitionis sensitivae*, study of sensitive knowledge, or the science of the knowledge we acquire by means of the senses. However, mainstream studies reduced it to dealing with other issues of greater social demand at the time of the art market expansion and interests, namely, to establish criteria for assessing the value of artworks and to set up an objective basis for the judgment of beauty. These constricted approaches have neglected concerns that rose since the dawn of Western philosophy, when Pythagoras related sounds and musical harmony with mathematical proportions and the golden section as proof of the perfection of the cosmos, or when Plato so accurately inferred the deceit and illusion of images and the potential for manipulation of passions by poetry.

Given the ambiguity of the term aesthetics, not only in everyday language but also in specialized texts on this topic, we must comply with a protocol that seems to be a life sentence imposed upon any research in aesthetics: the operational definition of the concept. Although authors such as Thornhill (2003, 9) deny that the domain of aesthetics can be defined (as philosophers before him have, Morris Weitz [1987, 153] for example), and intends to work just the traditional topics of aesthetics from an evolutionary model, the truth is that many of the problems that the emerging Darwinian aesthetics are inheriting result also, again, from the vagueness of the term and its implications. Works of great erudition get ensnared for not complying with this requirement, as the imprecision they tolerate in the concept places before them dangerous traps by the opportunistic use of its various meanings (metaphorical or literal, evaluative, descriptive or prescriptive) often contradictory.

These theoretical and terminological problems are common to different Darwinist works where the term of "aesthetics" keeps metamorphosing to denote preference, pleasure, art, decoration, good or bad taste, perception, fashion, style and beauty. Evolutionary aesthetics decides to start its way from David Mandel's 1967 pioneer work *Changing Art, Changing Man.*[2] A variety of approaches have since been proposed: experiments in neurological visual perception of color and form (Zeki 1980), the acknowledgment of non-western artifacts, customs, crafts, and rituals among native groups as art (Dissanayake 1982, 1996, 2007), surveys of preferences in parks and landscapes (Orians and Heerwagen 1992; Kaplan 1992) and in artifacts (Voland

and Grammer 2003), the moving body (Grammer and Thornhill 1994), reactions among infants to female faces in photographs (Etcoff 2000), aesthetic pleasure (Welsch 2004), the evolution of artistic creation and imagination (Velez Caicedo 2008), and the idea of art as an instinct. (Dutton 2009)

The problem is not the variety of topics that Darwinian aesthetics addresses. On the contrary, it is indeed involved in a wide spectrum of phenomena that go beyond established notions as "the genius of the artist" or "the ontology of beauty." The difficulty lies in the alteration of meaning in each case, as the term keeps slipping and camouflaging for the sake of the argument. To counter the uncertainty that Thornhill tolerates, it is necessary to determine its definition and hold on to its simple etymological meaning of aesthesis, derived from the Greek *aisthenastai* that denotes perception. From such perception, all the rest evolves: attraction, valuation, appreciation, fascination, interpretation, creation, and contemplation, since it is their very condition of possibility. We begin this exploration with our compass aligned on αισθητικός, what relates to sensibility. Aesthesis is receptivity, openness to the environment, the sentient and sensorial activity in any scale.[3] Without perception, there can be no artistic expression, nor appreciation of the graceful or tragic, and no beauty.[4] Not only Beethoven and Rembrandt have sensibility, dragonflies and bacteria also do.

Two tasks require particularly urgent attention if we are going to engage in an evolutionary approach to aesthetics: on one hand to decipher the different senses of aesthesis and on the other to remove the superstitions which have ossified discourse in this discipline. I specifically refer to the following tenets:

1. Culture is opposed to nature
2. Beauty is opposed to the useful
3. Aesthetic phenomena are purely cultural
4. The contemplation of beauty is disinterested
5. Aesthetic experience depends upon aspects and qualities of objects (proportion, order, rhythm, symmetry)
6. The enigma of beauty is solved only through philosophical analysis
7. Aesthetic judgments need objective criteria
8. Beauty is an exclusively spiritual value
9. Aesthetic appreciation is distanced
10. Any discussion of aesthetics must be conducted in relation to art and beauty

The aesthetic compass we'll use to guide us has a trembling little needle that points to a very different direction from conventional art-centric and ontological aesthetics with its dogmas. It also differs from geno-centric Darwinian aesthetics and its obsession with fitness and concentrates on those "little items" mentioned by James in the epigraph here, as well as on Darwin's "small trifling particulars."

SMALL TRIFLING PARTICULARS

> "I remember well the time when the thought of the eye made me cold all over, but I have got over this stage of the complaint, and now small trifling particulars of structure often make me feel uncomfortable."
> (Darwin to Asa Gray Apr. 3, 1860)

By adjusting small details in measuring the movement of the stars, Copernicus removed the earth from the center of the planetary system to place the sun in its place. Kant reverses the external location of space-time to locate it as internal a priori condition of possibility of knowledge within the transcendental subject. Feuerbach inverted the creation of man by God to God as creation of man. Marx overturned wealth and puts instead the workforce and plusvalue at the center of political economy to understand capitalism by observing details such as the quantity of shirts produced in a day, profit in sale, and the worker's salary. Darwin and Wallace inverted the theological version of God's creation of nature and placed instead natural selection of those small trifling particulars at the center and origin of evolution of the species. Decades later, Einstein takes the force of attraction of the mass and places instead curved space-time at the center of gravitational phenomena. It all seems as simple as turning an hourglass upside down, but it really took a tremendous effort to whirl against our stubborn mental inertia.

Darwin found the devil in the details. If one had to explain the evolution of the species, the starting point had to be in these "small trifling particulars" of everyday life, be it cricket elytra, the rubbing of a petiole, a woodpecker's beak, or the gesture of a monkey. It meant yielding our metaphysical musings for the attentive observation of these trivial details. Cyril Aydon (2003, 286), Darwin's biographer, points out his "almost superhuman ability to see things that other people did not notice. His powers of observation were as different from the average person's, as a hawk's are from a mole's. He also had a quite breathtaking ability to see, not only the thing itself, but its significance."

This observation of details brought forth the concept of "natural selection" that would be as fruitful to biology as "plusvalue" was to political economy. The animal origin of humankind and the exploitative nature of capitalism came to light in the nineteenth century by trying to explain the social and natural, but with a very dissimilar outcome: we now understand nature much better but the economy much worse (and we are ruining them both).

The parallel between Marx and Darwin is reinforced: both are anathema to the Church. Darwinism was popularized and demonized by the cliché of seeing it as a voracious survival of the fittest in nineteenth-century Spencerian social Darwinism and twentieth century eugenics. It must be emphasized, however, that Darwin never used the idea of "survival of the fittest" to support or justify the fittest as the most ferocious, predatory, or aggressive but instead stressed that society must have means to protect the weaker and helpless. In turn, vulgar Marxism reduces human beings to mere labor agents under the inexorable laws of capital. Hence, discourse in both Marxism and Darwinism is governed by economy because at the bottom of human and animal endeavors always resides the basic need to get the material means to survive.

Such parallelism is suspended when it comes to the aesthetic because while a huge number of texts on aesthetics proliferated during the first half of the twentieth century from a Marxist approach, this was not the case with Darwinism. No attention was paid to Darwin's writings on the sense of beauty in animals since art and beauty were taken to be elevated and lofty topics, whereas the evolution of animal species and their sexuality was considered an unrefined and even vulgar subject matter.

The Vulgarity of the Sublime

Aesthetics and art have sufficiently proven their great value and expediency for political propaganda. The USSR and its satellites, communist party members and sympathizers committed their artists and intellectuals to produce propaganda for the service of the socialist State. Gramcsi urged "the organic intellectuals" to perform the task of confectioning the ideology for this proletarian revolution. It seemed that Marxism could finally ground abstract Hegelian speculation and use aesthetics as a valuable tool in the ideological and practical liberation of the proletariat. In this Marxist utopia, both the prospective (vision of the future) and the prescriptive (normative) dominated the agenda. Artistic production pictured itself from a role of leadership in pioneering this great social, cultural, and economic revolution. Socialist aesthetics, with György Lukács as a major figure and his followers at the Frankfurt School, pointed to this direction with fervor, encouraging art to fulfill its emancipatory duty. Sartre urged the intellectual to commit himself to the cause, Althusser defined the ideological apparati of state in their task of interpelating the subject as subject-of and as subjected-to such state. These early decades of the twentieth century seemed exciting times, with the promise of a useful aesthetics marching on the right side of history that could now transform reality, not just contemplate it.

But aesthetics helped develop propaganda more than it developed theory and legitimized the aesthetician and art critic in their role of ideological watchdogs of good bourgeois taste and anti-bourgeois good conscience. What Marxism understood as "art" was, disastrously, a purely capitalist bourgeois, elitist phenomenon, a dilemma that Lukács was forced to face when he proposed nineteenth century bourgeois art as a model for twentieth-century proletarian society. In retaliation, the same bourgeois art became the pompous Stalinist art that would end up crushed under the blocks of the crumbling Berlin Wall until it became nostalgic memorabilia of the communist era. Today this style is resuscitated in the sarcastic conceptualist homage by Komar and Melamid with ridiculously solemn portraits of Stalin and comrades.

As prophesied by Marx, the internal contradictions of capitalism that burst with the cardiac attack at the heart of Wall Street's worst crisis during 2008–2009 ended up exhibiting that it is indeed just a castle made out of cards: the stocks. It is surprising that a system founded on what is most concrete and banal, namely money, is in fact being supported by the most abstract and imponderable: faith (in stock exchange). The point I want to emphasize is that while Marxist aesthetics were prolific, Darwinian aesthetics remained paraplegic. Just recently some texts on this subject are beginning to be published with a delay of more than a century that try to explain art and taste as adaptive mechanisms that optimize survival and reproduction, as discussed in the following sections.

Beyond their vulgar versions, Marxism and Darwinism still have something else in common: both lead to reflect on aesthetics from a functional and practical perspective, in one case as adaptive mechanism for survival and in the other as an aid to class-consciousness or alienation, including emancipation (Adorno). Both require observing structures and behaviors of everyday life and both take the human being as a creature of flesh and blood that acts to fulfill specific purposes such as its needs for food, shelter, and companionship, care for offspring, desire for status, fun, and affection. But while Marxist aesthetics was left behind by failing to obtain empirical evidence or explanatory results and remains in its current ideological position, Darwinian aesthetics, I will contend, has significant explanatory potential just being developed.

It may not be correct to reveal family secrets, but the many similarities between Marxism and Darwinism already turn into an open secret that, besides their obvious family resemblances with utilitarianism, they procreated together a natural legitimate son: pragmatism.

The Body in Theory

Marx and Darwin allowed the in-corporation of the corporeal into theory, if I may be so redundant. The force of labor for Marx is the worker's body, what is exploited is his very life and fiber, his corporeal condition coagulating human vital energy in commodities. In turn, Darwin began a genealogy of the body because evolution is nothing other than the development of corporeality.[5] In line with Hobbes's view of *homo homini lupus*, vulgar Darwinism was translated as the "survival of the fittest," a law of the jungle. But who is the "fittest"? The most aggressive? The richest? The most prolific? The most beautiful perhaps?

Alfred Lord Tennyson's phrase "red in tooth and claw" often quoted among Darwinists, is used as a metaphor for the cruelty inherent in nature by predators' tooth and claw dripping with their victims' blood. This sense of ferocity has been exacerbated in Richard Dawkins' (1976) "selfish gene" theory of quasi-mechanical dynamics in gene replication. For Dawkins, evolution is a mechanical consequence of blind genetic replicators that utilize bodies as vehicles best suited to guarantee their replication in various ecological niches. Bodies are just gene replicating machines. One wonders then why so much effort is spent into producing phenotypes when that energy could be used more efficiently in the sole multiplication of naked genes without wasting resources on excessive devices such as bodies. This waste may be explained perhaps by a Hegelian curiosity of sorts in objectifying for the sake of contemplating its own possibilities and contradicting Dawkins. In other words, a blind and insatiable replicating machine does not seem to be in operation but on the contrary, a subject stubborn to see the world and play by trial and error all possibilities of being in this world with the skill of an expert gambler. This individual subject fits its senses to see better, hear better, and play better, like the wolf in *Little Red Riding Hood* would have said.

Such splurging throughout a variety of species and prodigious forms seems to violate the law of entropy, because instead of tending to homogeneity and disorder, evolution tends to diversify and differentiate. One suspects that what is at stake in nature is a perspicacity to integrate and blend the simple to greater complexity and to peek into the unknown. Since the binding of quarks to form hadrons into a variety of atoms that develop into inorganic and organic molecules that enable replicating DNA, and from there the shuffling and combinatorial diploid eukaryotic organisms up to sexual preference selection and the generation of varied biomes with multifarious interactive webs, all seem to point to the possibility of a God who would create the world out of mere curiosity. And God saw that it was so interesting . . . and got excited.

THE SIGHT OF A FEATHER IN A PEACOCK'S TAIL, WHENEVER I GAZE AT IT, MAKES ME SICK

The enigma of the peacock, a singular aesthetic and absolutely excessive event in nature, was so enigmatic that it literally made Darwin sick, as he confesses in a letter to his friend Asa Gray on April 3, 1860 by the quote above. No wonder. This magnificent peacock tail cast doubt on random mutation and natural selection as the sole explanatory principle of evolution in *The Origin of Species* that predicts that a peacock with a short tail would have been selected over one with the long tail simply because it is more practical for survival. So hard to maintain and show off, so inconvenient in making its owner more conspicuous to predators, slower to escape danger, more in need of nutrients, and more vulnerable to parasites exposing its flaws to females, this huge tail did not seem to find a coherent explanation in the Darwinian theory. It was all perfectly in place before he realized this anomaly. Such extravagance became like a ghost that haunted Darwin's paradigm threatening it to collapse.

Darwin's sickness turned into a real passion for explaining it. Despite the criticisms and objections even from those who could help him solve it, as his co-author Alfred R. Wallace, Darwin assumed this enormous challenge notwithstanding its great intellectual cost: the effort to write another book, *The Descent of Man and Selection in Relation to Sex*, almost double the size of the *Origin of the Species* (there are 502 pages in *Origin* and 899 in the *Descent*), and the punishment of having to remain almost ignored by academic publishing for a century. In this second text, Darwin confesses that he collected notes on the origin of man with the intention of not publishing them, as merely the slight mention in the *Origin of Species* that "light will be thrown on the origin of man and his history" (Darwin 1859, 254) caused such commotion as to discourage anyone. He nevertheless took the study of natural selection in *The Origin* and of sexual selection in *Descent*.

By the new version, the process of evolution is due not only to the blind and fierce mechanism of natural selection of the fittest by random mutations and selective retention of traits in the struggle for survival, but also to something different and more radical: the idea that the female of each species could be running the selection process. To top it off, this is accomplished through aesthetic criteria, the superfluous almost by definition. Evolution at the hands of the aesthetic whim of females!

This conclusion reflects Darwin's intellectual honesty considering his own misogynist bias prevalent in the Victorian context (as if Queen Victoria where not a female), and had the bad taste to write that: "The chief distinction in the intellectual powers of the two sexes is shewn by man's attaining to a higher eminence, in whatever he takes up, than can woman—whether requiring deep thought, reason, or imagination, or merely the use of the senses and hands" (Darwin 1882, 564). So the eternal feminine now billed Darwin's prejudice dearly: again as Eve, Lilith, Pandora, Helena of Troy, Cleopatra, and Malintzin, it was entirely the female's fault.

Darwin was ridiculed for his idea of female selection and even in 1960, as Trivers (1985, 333, 336) notes, an explanation was taken seriously according to which females were wooed not because they could choose partners but because they were too lazy to mate naturally and were afraid of being touched since when a predator touches them, they die. Such a theory is false as proven by the highly selective sense of females in various species. For instance, Michael Ryan's (1998) experiment with *Physalaemus postulosus* frogs proved that they are able to accurately distinguish the size of the male by the simple croaking tone and therefore select the largest.

The female is at the helm of the evolution of multiple species as she requires to be captivated by the male specimen whose particular features she chooses to pass on to the next generation. In many cases, she not even has to wait to be seduced, but goes straight to the male that is most attractive to her and copulates with him. The selection of such exotic luxuries of nature as birds of paradise, pheasants, and peacocks have no other explanation than, say, this "aesthetic" taste of females in total rebellion against evolutionary instrumentalism. The peacock phenomenon is a challenge not only to misogyny but to the pragmatic heart and marrow of evolution, because choosing the beautiful rather than useful requires some explanation. We owe to the females of each species the variety of colors, shapes, and ornaments of nature by selecting and cultivating the finest for reproduction.

> Just as man can give beauty, according to his standard of taste, to his male poultry, or more strictly can modify the beauty originally acquired by the parent species, can give to the Sebright bantam a new and elegant plumage, an erect and peculiar carriage—so it appears that female birds in a state of nature, have by a long selection of the more attractive males, added to their beauty or other attractive qualities. No doubt this implies powers of discrimination and taste on the part of the female which will at first appear extremely improbable; but by the facts to be adduced hereafter, I hope to be able to shew that the females actually have these powers. (Darwin 1882, 211)

This female frivolity implies, therefore, that what is at stake is not only direct instrumental criteria, but the aesthetic as well. This is a scandal that not only upsets the misogynist bias when it recognizes that females drive the evolution of certain species but also puts into question the evolutionary formula of "blind mutation and natural selection" to the opposite: a deliberate and very discerning mode of selection. Darwinian functionalism derives, paradoxically, into hedonism and caprice.

During the nineteen twenties, Fisher (1999) proposed an answer to this enigma when he coined the idea of the "runaway process," a hypothesis that assumes that not only traits but preferences are inherited and thus traits that are preferred have an advantage in selection. Fisher explains the case of the peacock as a result of preferences in females inherited by their daughters, and traits in males inherited by their male offspring, who will be the preferred for mating.

Females in many species are not forced to mate with the bravest male winning all contests at the birds' public square or lek, but seduced by the most charming. "The rock-thrush of Guiana, birds of Paradise, and some others, congregate; and successive males display their gorgeous plumage and perform strange antics before the females, which standing by as spectators, at last choose the most attractive partner." (Darwin 1859, 89) He adds that "the exertion of some choice on the part of the female seems a law almost as general as the eagerness of the male." (Darwin 1882, 222) By "eagerness" Darwin is implying males' low discriminatory power, an understatement to denote how easy men can be.

This explanation brings up another puzzle even more difficult to solve: Why do females require beauty to mate? Do they feel pleasure at the sight of a male peacock's tail? How important is the beauty of the male to a female if she stays away from him immediately after copulation anyway, as is the case of polygamous peacocks? Is the female cricket moved by listening to the stridulating music of the male? Does she really interpret it as something close to "beautiful" or something else? Does the peahen admire colors and proportions or rather perform a calculation by phenotype of resistance to parasites and genotype quality indexes? Nagel asked, "What is it like to be a bat?" To solve this mystery, I would really like to know what it like is to be a peahen.

Stripping the argument:

1. There is evidence on the preference of certain traits over others in some species that do not appear to relate directly to any useful purpose.
2. There are species that contradict the law of natural selection in that they are focused on the reproduction not precisely of functional fitness.
3. To our knowledge, we have no proof nor can we be sure that there is no sense of "beauty" in other species, but that their preferences for vivid colors, symmetry, proportion are consistent with human aesthetic evaluation criteria is a fact.

Darwin emphasizes the evolution by sexual selection of "their courage and pugnacity—their various ornaments—their contrivances for producing vocal or instrumental music—and their glands for emitting odors, most of these latter structures serving only to allure or excite the female." (Darwin 1882, 210–211) The main consequence of this approach is that the evolution of creatures appears not to be blind at all but very sharp, sensual, and selective to the extent that by contributing to it we are now rewarded with the experience of beauty (whatever that means), as well as alerted by the sense of ugliness.

God could be a bad mathematician when She calculated the origin of the world at 5775 years ago instead of 13.73 billion years, an error of only 7 zeroes. Nevertheless, by the biblical command to Abraham, "Be fruitful and multiply," God proved to be an excellent Darwinian. She is also a prodigious aesthetician by electing the female sensitivity to beauty as the guiding direction through much of evolution.

PREDICTING THE PLEISTOCENE

The most unpredictable fact of the Pleistocene was the emergence of an apparently unique phenomenon in the universe: the human mind, a faculty capable of predicting the future from the past. This ability transforms the australopithecine hominid to Homo habilis when he manages to predict that by striking a particular stone with another of particular consistency at a particular angle, one of the two would break in a particular way. Predicting the Pleistocene involves predicting the material specificity of each stone and the approximate place of a crack, and is extended even to the prediction of the route taken by herds, the arrival of spring by observing the flight of birds, and the meaning of a trace upon the ice, mud, grass, or sand. The Pleistocene mind can even predict the direction of wind, the resistance of water, the duration of provisions when 60 thousand years ago it performed the incredible feat of building a vessel to navigate from Southeast Asia to Australia. The Pleistocene ended 10 kiloyears ago, with the heat of the current Holocene, but its consequences are lived and felt by us every single moment.

Evolutionary psychology assumes that our choices and reactions are a product of the mental and anatomical conformation that our ancestors acquired 2 million years ago when they roamed the savannah as nomadic hunters and gatherers until we evolved into homo sapiens. This position is in line with the premise that there is a universal human nature which is a product of evolution of our physiological and psychological mechanisms as adaptations to respond to problems of survival and reproduction in the Pleistocene. (Cf. Barkow et al. 1995) We should therefore recognize that we partly contain, as Russian matrioshka dolls, homo ergaster, erectus, habilis, hominids, primates, mammals, vertebrates, metazoids, eukaryotes, and prokaryotes from whom we descended. The origin and development of human adaptation to living conditions in this stage determined the morphological stability that we gradually acquired. Are we aware that already two hundred thousand years ago there was a creature that played the flute, learned to sew, and painted animal and human figures with great expressivity? I wish I could hear her playing such a flute. (Bahn and Pettit 2013, 5)

Studying the organism in reverse as a set of adaptations over millions of generations is the strategy that Darwin used for explaining the mechanics of evolution. Certain traits and organs are formed by infinitesimal changes until structures emerge as spectacular as the spherical eye lens of the fish and fly, or the heightened canine sense of smell able to perceive traces left by others not only in space but in time, as well as Broca and Wernicke's areas in the neurocortex that enable the coming out of sophisticated human language.

The transformation of the phenotype is due not only to random genetic mutations and selective retention of favorable traits, according to the Darwinian formula, but to environmental impacts that can cause changes without mutations in the development of an organism simply by activating dormant genetic traits of the genotype that reveal genetic variants already present in a population. These variants are captured later by natural selection to be reshuffled in sexual reproduction whose combinations result in more effective survival phenotypes. The fact that changes induced during cell development may become hereditary involves an interaction between the epigenetic level (epi, around) and the genes which can affect each other in both directions. Therefore, the variation of species depends not only on mutations but also on environmental changes that accelerate evolution to produce different phenotypes from the same genome.

Predicting the Pleistocene, then, is not only the oxymoron of predicting the past but a crucial issue during a period of highly unpredictable weather extremes from heat and drought to glaciations. Those who could detect dew drops on leaves or stones in the desert, discern the resistance of ice on a lake, and quickly distinguish the fur of a bear or tiger through a bush's leaves, the scales of a reptile among the weeds, or ripe fruit from a very long distance passed not only their own discernment but their life to their descendants. Perhaps the origin of our passion for gold and precious stones, our preferences for the color red and animals' fur may be found in our ancestors' experience in life or death situations that depended on these perceptions.

The Savannah Hypothesis

> By Hera, a fair resting-place, full of summer sounds and scents. Here is this lofty and spreading plane-tree, and the agnus cast us high and clustering, in the fullest blossom and the greatest fragrance; and the stream which flows beneath the plane-tree is deliciously cold to the feet. Judging from the ornaments and images, this must be a spot sacred to Achelous and the Nymphs. How delightful is the breeze:—so very sweet; and there is a sound in the air shrill and summerlike which makes answer to the chorus of the cicadae. But the greatest charm of all is the grass, like a pillow gently sloping to the head. My dear Phaedrus, you have been an admirable guide.
>
> —Plato *Phaedrus* [230 b–c]

Savannah is the classical image of paradise with a lake surrounded by green grassland, trees, a water stream, and a mountain at the distance shrouded by clouds in bucolic scenes of happiness. It is the favorite background in paintings expressing well being from the Renaissance on, since it is the most harmonious pictorial space for the figures in the foreground. Such background was Bosch's choice for *Garden of Delights*, Leonardo's for *Gioconda* and *Virgin of the Rocks*, Raphael's for *Madonna of Goldfinch*, El Greco's for *View of Toledo*, as well as Titian's for *Bacchus and Ariadne*, Bruegel's *Return of the Herd*, and so on, the pleasant and idyllic place that is repeatedly evoked in culture.

Appleton (1996) investigated preferences for certain landscapes, proposing the principle of *prospect-refuge* for the perfect site to explain the human predilection for spaces that offer a wide visual horizon with lots of information as well as areas to hide from predators. In this light and based on results of surveys on landscape preferences, Orians and Heerwagen (1992) suggest the "savannah hypothesis" by which hominids who preferred the savannahs were fittest in natural selection and passed on to their descendants this preference as the most favorable and stable environment. Parks and gardens effectively reproduce this human-friendly environment and our preference seems to confirm such hypothesis. Experiments on landscape preferences performed by Kaplan (1992), in a variant from Appleton's, advocate categories of mystery and coherence as aesthetic aspects for evaluation. He establishes two criteria for human preference for landscapes: exploration (in relation to the complexity and mystery that may store information) and understanding (in relation to the coherence and legibility of the landscape).

With wit and sense of humor Vitaly Komar and Alexander Melamid (1999) hired a poll survey company to find out the tastes in painting among Americans and other nationalities in 1994–1997. The results obtained by telephone in a first sample of a thousand and one adults are poured into a painting that collects, obeys, and expresses the popular taste, a truly democratic painting and quick gain for art dealers. The favorite landscape of respondents overwhelmingly turned out to be the prairie or savannah. They were able to raise funds to expand surveys across various countries such as China, France, Finland, Iceland, and Portugal, and again found this preference for more or less the same blue landscape (except in Holland, the birthplace of Rembrandt, Vermeer, and Van Gogh where respondents preferred an abstract painting!).

We could have inherited our preference for blue from our primate relatives because, according to Humphrey (quoted by Sebeok 1979, 39), in experiments monkeys' chromatic order of preference is blue, green, yellow, orange, and red. Moreover, according to his results in image preferences using photographs, their favorites were, in this order: other animals / apes / men (caregiver) / flowers / abstract painting / meal. Yes, they prefer abstract art to food!

This double mockery of the survey method and of democracy in the artworld is taken seriously by some authors as evidence for the Pleistocene savannah hypothesis. (Dissanayake 1998; Dutton 2009) The fallacy of this extrapolation is that such a test was designed to find out tastes in painting, not in landscapes. Therefore, the sample is contaminated by conventions of representation and illustration and by the widespread frequency in the reproduction of certain types of scenery in a given cultural context.

In the case of direct preferences for landscapes, surveys conducted in the Netherlands and Sweden question the predilections for the savannah and are more inclined to prefer forests. In fairy tales and in romantic and chivalry literature forests are represented as places of mystery and danger. The wolf in *Little Red Riding Hood* appears in the forest and Hansel and Gretel are lost, precisely, in a forest. Also the Hobbit and Harry Potter had their adventures in the woods. Medieval knights in epic poems as *El Cid* or *Karel Elegast* occur in forests, and trees coming to attack Macbeth move from forests. Lady Chatterley is in the woods with her lover, the forest keeper. But there are also images of the jungle in Rudyard Kipling or in Jorge Isaac who describes the mangrove landscape; as each provokes different connotations and evocations.

Hence not only the savannah and forest are at the top landscape preferences, especially if we take into account the modern phenomenon of mass migration to the beach at every opportunity. One can speculate that this occurrence could be attributed to our pre-hominid or protozoan stage that longs to return to the sea or, in the opposite direction, to our amphibious stage searching for the beach from the sea towards the mammalization process. The answer, however, is so obvious that it is not even worth mentioning it except as an undeniable confirmation of the Darwinian thesis: the attraction for the beach is less in the contemplation of the sea landscape than of the animals' torso, biceps, waist, chest, and buttocks parading around. The beach is the ideal place for sexual selection in line with the most orthodox evolutionary psychology.

Aesthetic Adaptations

Jerome Barkow, Leda Cosmides, and John Tooby (1992, 8–9) propose a basic distinction between two key terms of Darwinism, "adaptation" and "adaptive" and note that "an *adaptive problem* is a problem whose solution can affect reproduction, however distally. Avoiding predation, choosing nutritious food, finding a mate, and communicating with others are examples of adaptive problems that our hominid ancestors would have faced."

The *adaptive* implies a particular purpose and is focused on the future, contrary to adaptations that result from the past and are read as evidence of a process of interaction with the environment. In the argument that Stephen Jay Gould and Richard Lewontin make against Edward Wilson and Richard Dawkins's neo-Darwinism, the main quid of debate is located in this relation to the environment. Gould and Lewontin argue that organisms not only adapt to their environment but are also part of it and transform it, all within a multiway interactional dynamic between organisms and others that constitute their environment. (Lewontin et al. 1985; Gould 1996; Gould and Lewontin 1979) Such dynamic interaction has been moreover reinforced by the advances in our understanding of epigenetic processes on the activation or silencing of genes by context effects. The environment influences the development, and has a key role in determining which genetic variants are selected in the organism's shape. As Jablonka and Lamb (2005, 286) have pointed out, a learned response to the environment can become an innate behavior. We do not yet know how.

Adaptation, on the other hand, is the key to evolutionary theory and it is as important to the Darwinian paradigm as the concept of "commodities" for the Marxist paradigm, because each adaptation can be read as an index of the evolutionary processes that have shaped it. Thornhill (2003, 13) defines an adaptation as "a phenotypic feature that is so precisely organized for some apparent purpose that chance cannot be the explanation of the feature's existence." An adaptation is the effect of a response to material selection, where the selection is defined as a non-random differential reproduction of individuals. Every organism is an integrated web of like phenotype adaptations to survive and reproduce. The body of a creature can be deciphered as a map of the environmental pressures it confronted throughout its evolution.

Adaptiveness, conversely, is not a criterion in the evolutionary paradigm. According to Thornhill (2003, 341) the only criterion for understanding evolved adaptation is functional design. Therefore, each adaptation is a physiological and cumulative memory of the past forces and choices that shaped it. There are adaptive adaptations (which help to integrate the organism to the environment) and maladaptive adaptations as pointed out by Boyd and Richerson (2005, chapter 5).

The point is that "each species has a species-specific psychological machinery designed for aesthetic appreciation of species-specific features of the environment that impacted reproductive success in each species' evolutionary history." (Thornhill 2003) But why precisely does Thornhill characterize it as "aesthetic"? Would it not be sufficient to say "designed for the calculation of usefulness of environmental aspects"? Thornhill (2003, 13) believes that "the psychological adaptation causally underlies *all* human feelings, emotion, arousal, creativity, learning and behavior" and assumes that these adaptations are always defined by fitness. Consequently, for the author, perception of symmetry, harmony, truth, unity, and order have a specific purpose in sexual, social, or environmental selection, rather than merely for contemplation.

Thornhill proposes ten categories of human psychological adaptation for aesthetic valuation: 1) of landscape features, 2) of non human animals, 3) of acoustic behavior of non-human animals, 4) arising from daily or environmental cues that signal a need to change behavior, 5) of human bodily form,

6) of status cues, 7) of social scenarios, 8) based on skill, 9) of food, and 10) judgments of ideas. (Thornhill 2003, 27–31) As with the Chinese encyclopedia mentioned by Borges (see the section "The Perfect Taxonomy" in chapter 2), it is difficult to know what criteria operates in this classification, what is meant by "acoustic behavior," what is the difference between 1 and 4, why he doesn't include movement, animals' body language, smell, ornamentation, and other aspects of sensory valuation that were so important to Darwin. Moreover, the main problem with this characterization is anthropocentrism, inconsistent with a Darwinian perspective, and which I attribute to the vagueness of the term "aesthetics" in this flirtation with traditional notions of beauty and art.

Let us address this terminological ambiguity at once. Sensibility is displayed in many activities of everyday life, as it is not dedicated only to entertainment and amusement through the arts. The relationship between what I termed "prosaics" (Mandoki 1994), and "poetics" (aesthetics of art) is not one of opposition but of inclusion, the latter within the former as a very specialized manner of aesthetic activity.[6] Both are part of anthropo-aesthetics which in turn are inscribed, as we shall see, within zoo-aesthetics and bio-aesthetics. What concerns us in this work is bio-aesthetics as the study of sensibility through various creatures intertwined in their environment affecting their perception and behavior at different scales.

Therefore to explore aesthetics from an evolutionary perspective we should be able to distinguish between art and aesthetics, between artistic production and aesthetic reception, between social dimensions and cultural matrices, between socio-aesthetics and bio-aesthetics and continue the thread of aesthetics from the other side, not precisely Duchamp's *Fountain* or Yves Klein's *Vacuum* but through the jellyfish and starfish in their sensitivity to temperature, water orientation, and light. Not too bad, is it?

AN EXHILARATING GAME OF NATURAL SELECTION IN THE FIELD OF EVOLUTION

In This Corner

This thrilling competition for The Cup of Truth between teams of scholars trying to explain the relation of nature and culture can keep us at the very edge of our seat. The brawler is genetic determinism, initiated by George C. Williams, followed by John Maynard Smith and contagiously spread by the selfish meme of Richard Dawkins's "selfish gene." With the underpinnings of William Hamilton, Robert Trivers, and Edward O. Wilson's excellent displays, the meme itself is confirmed by its own spectacular propagation. According to this position, cultural memes are replicators that jump from one brain to another, described by Richard Dawkins (1976) as "tunes, ideas, phrases, fashions, ways of making pots or of building arches." Cultural replicators, as genetic replicators, also require longevity, fecundity, and fidelity as conditions for replication. A book has more replication power than the spoken word because of its greater longevity and ability to transcend to several generations; it has higher fertility to the extent that new ideas generate from books that are republished or reread, and it has higher fidelity when its concepts are reproduced with relative accuracy. But the oral word, if eloquent, is more contagious than the written in a process which is more akin to viral rather than genetic replication.

The meme metaphor refers to the propagation or contagion of fads such as miniskirts, piercing, unicorn tattoos, the emoticon smile, karaoke, the "yes we can" motto, or hip hop hands' gesture as well as politically correct clichés. As a theoretical term, however, the meme has too many problems because of its atomistic conception (similar to the "smallest unit of meaning" that continental semiotics was vainly pursuing) as it ignores the context and pretends to define Euclidean figures when in fact we are dealing with a dense reticular web pregnant with resonances, overtones, connotations, derivations, reconstructions, and combinations of actions and meanings.

This genocentric team thought it would conquer the pinnacle of the evolutionary explanation of culture with a single player named "the meme" and in a single run, the gene, and threw it upon the field of humanities and social sciences.

The sociobiological approach starts from Ronald Fisher and John Burdon Haldane and crystallizes in what has been called the New Synthesis that integrates Darwin's theory of natural selection with Mendel's population genetics. This approach caused great controversy within the scientific community for attempting to apply "the systematic study of the biological basis of all social behavior." (Wilson 1980, 4) Wilson developed a comprehensive compilation of observations for the study of patterns of social behavior from ethologists, zoologists, and behavioral psychologists trying to understand how collectivities among animals in various species are organized. In the last chapter of his *Sociobiology*, he tries to apply these observations to human societies and culture under the premise that "genes hold culture on a leash." As a result he was insulted, accused of being racist, eugenicist, and fascist, his lectures were sabotaged, and he was even physically assaulted when someone dumped a pitcher of water over his head during one of his lectures.[7] Here we have an animal behavior to combat a theory and by doing so unintentionally almost proving its point.

Cosmides and Tooby (1992, 61) criticize social constructionism (cf. Berger and Luckman 1966) implying that culture lacks any autonomy over genetic determinism. They propose the distinction between "epidemiological culture" and "evoked culture" in an attempt to leave a place, even if minimal, to the social context through the notion that culture is evoked by the environment. What exactly, however, they mean by "evocation" is far from clear. Along this line, Dan Sperber proposed to study culture from an epidemiological approach as a phenomenon of distribution, despite of the fact that, by definition, culture is always already epidemiological. The problem in applying Sperber's approach begins with the dubious dichotomy and categories of "mental representations" and "public representations" as if the mental was not already structured by the social, starting with language, and vice versa, as if public representations were not necessarily mediated by the mental.[8] Moreover, the concept of "representation" is also quite problematic, as Varela, Thompson, and Rosch (1992) pointed out.

Furthermore, Darwinists like Trivers (1971) have the nerve to apply concepts from the New York Stock Exchange (investment, cost-benefit calculation, credit) to the much idealized and sacrosanct parent-child relation. But we cannot deny the fact that in the human species the mother invests significantly more energy, time, and material resources on her offspring than the father does. An ovule is metabolically much more expensive than a sperm, and this biological fact has been one of the conditions of laboral, religious, and family exploitation of women.

Although it is tempting to compress everything into a single formula of gene replication, we can not ignore differences in level, scales, ramifications, and qualities. Genes are not only material molecules formed in strands of DNA but replicable information; cells are not only matter and form and replicable information but also intentionality and motility. Dawkins's genocentrism could be understood as a voracious pansemiotic adventure of focusing entirely on genes in terms of information that use the body as a machine to replicate. He could consequently be considered an involuntary semiologist taking organisms as semiotic packages of genes diligently devoted to deliver them or, better, impose them on other organisms. What is not in doubt is that evolution itself is a process of widespread semiosis, since what is passed on from one organism to another through inheritance is nothing other than information. And this information is ever more finely diversified.

For Brian Goodwin (1996, chapter 4) genes are dances in the morphospace or instructions for deploying bodies or matter in space. He insists that evolution depends on conformational structures according to certain mathematical and physical laws that determine morphogenesis in embryology. He questions the red in tooth and claw version of evolutionary fight and competition, or calculation of costs and benefits, and proposes instead that the current postmodern world may think of a plurality of paradigms for a more qualitative-oriented health, creativity, relationships, and lifestyle. However, in choice of paradigms a calculation of costs and benefits already comes into play since we would not choose a complex and baroque paradigm with poor explanatory power over another that is simple, heuristically fertile, and clear, as in Occam's razor. Goodwin caricatures Dawkins as an inverse fundamentalist Christian, but who preaches cynical selfishness under the creeds: 1) humankind is born in sin, 2) we have a selfish heritage, 3) we are doomed to a life of perpetual conflict and struggle, 4) but salvation comes when the fittest organisms get to maximize their fitness. Darwinian fitness for Dawkins, according to Goodwin, is a kind of Calvinist determinism in the race of life. Goodwin's criticism, however, misses the point because Dawkins's position in this particular case is not value laden. It is descriptive, not evaluative.

Daniel Dennett tries to perform a Copernican knowledge revolution in direct opposition to Kant's by eliminating the subject of sensibility from the epistemological horizon. He thus proposed a fully de-aesthetized vision of consciousness in contrast to Kant's Transcendental Aesthetics. At the risk of returning to naive realism against which Kant builds his massive philosophical architecture, Dennett loses not only the subject but also the world: in their place only hardly distinguishable and relatively random computer operations remain, as he reduces the brain to a flexible information device. For Dennett there is no qualia, no emotions, and no tastes of things, no feelings of pleasure or pain.[9] Thus, rather than the common reprimand by his critics against his book *Consciousness Explained* that it is really "Consciousness Explained Away," one might therefore entitle it "Experience Unexplained," the most readable anti-aesthetics text written until today, only to be toppled by a future text "experience un-experienced."

In the Other Corner

The criticism against neo-Darwinism and sociobiology has been triggered from all possible fronts: anthropology, paleontology, zoology, biology, genetics, sociology, and Marxism. For instance Lewontin and Gould have pointed out the ethical and political consequences that may result from genetic determinism of behavior by the sociobiological approach. As Lewontin (1985, chap. 5) holds, genetic determinism could lend itself to naturalize the fact that men beat their wives due to their increased level of testosterone; another consequence could be to restrict access to education for individuals with low IQ tests rating, supposedly "not to waste resources."

Gould is situated at the antipodes of Dawkins in evolutionary theory. Both acknowledge the huge difference of interpretation between them since the former proposes the concept of evolution as being not only discontinuous, but subject to catastrophes such as dinosaurs' extinction by a falling asteroid. Gould called "punctuated equilibria" the fact that a species is more or less stable and will not change progressively from one generation to another by infinitesimal changes, as implied in the formula of random mutations and selective retention. With Eldredge, Gould argues that not all species emerge gradually since there are sudden changes or revolutions, large-scale macro-evolutionary jumps to certain species. As a case in point he underlines the fact that in the fossil record there are no intermediate cases, no evidence of continuity of the species.

In the other corner of the debate, Marshall Sahlins (1976, 6–7) challenges that the driving force of evolution is the genotype's self-maximization. He points out that contrary to what sociobiology proposes, there is no necessary relation between social institutions and innate human tendencies such as aggression, altruism, sexuality, paternal or maternal protective attitudes. There is no evidence either, according to Sahlins, that there is a homology between biological patterns and social relations. Indeed, there may be no homology, but research regarding the evolutionary origin of emotions and of actions such as altruism, incest avoidance, and control of aggression, and of institutions such as monogamy and gender division of labor has rendered significant empirical evidence that demonstrates the resonances of human behavior with animal species.[10] For Sahlins culture is the result of symbolic production and the will of humans, but where or why such symbols emerge in general remains unexplained by his approach.

From a Marxist perspective, biological evolution ends when production or work begins. It establishes a radical dichotomy as if production and work did not already exist in other species, and as if work had no effect on biology and evolution. Thus social and cultural evolution are considered from this point of view as completely autonomous and distinct spheres from biology. This position is unconvincing given research on labor and social behavior of ants, termites, bees, or the political organization among primates.[11] We know also that the gene is at work when we observe the behavior of members of monarchies and oligarchies, as well as the jet set, political nepotism, and monopolization by certain families of material resources, bequeathing political and social capital to their progeny. These practices remind us of kin selection theory proposed by Haldane and Hamilton, since how else can we explain the seemingly irrational behavior of powerful individuals sacrificing their lives to amass such wealth far beyond what they could possibly enjoy in their lifetime, in order to bequeath it to descendants? (Haldane 1955; W. D. Hamilton 1963)

In Between

There are issues over which a middle ground is not possible, as in the use of bigoted social Darwinism in Galton and Gobineau's writings. Criticizing genocentrism from the left is essential to alert researchers about its political consequences, but it should not imply canceling research of polemic issues. It is not a question of imposing self-censorship by political correctness and prematurely aborting possibilities for reflection, knowledge, and emancipation, but theories must be assessed solely by their scientific correctness, not by their political correctness, and yet be acutely aware of their political relevance because the theoretical play is also a political play.

The evolutionary paradigm itself does not imply a right or left political stance, since it may be able to prove better than Marxism "the potential of early social environment in shaping the adult mind," as stated by Wright emphasizing the importance of promoting the best possible cultural conditions for all children. (Wright 1994, 8) In short, to arbitrarily impose the nature-culture dichotomy supposedly to save free will, as I understand Sahlins's position, is a useless and dangerous sacrifice. Useless because evolution enables a degree of freedom as in sexual selection, and dangerous because it imposes taboos, pre-set thought, and censorship that restrict research and thus hinder realistic and grounded solutions to pressing social problems.

Moreover, the "naturalistic fallacy" assumes, wrongly, the fact that if something exists or is natural it means that it is necessary or unchangeable. Disease is natural, but we must confront it and try to cure it. Therefore the fact that something is a product of natural selection does not mean it is immutable or desirable. In Wright's (1994, 31) words there is "no reason to adopt 'values' of natural selection as our own. But presumably if we want to pursue values that are at odds with natural selection's, we need to know what we are up against." Emotions such as altruism, compassion, empathy, and love have a biological origin, which does not take away or add them credit, but helps in guiding us for the good of all.

This is the background of the debate about Darwinism between cultural anthropology and evolutionary psychology in their common effort to explain human behavior. Both sides have stereotyped the opponent: the culturalists assume that culture is an autonomous phenomenon from biology but do not explain how and why it happens, emerges, develops, and shapes the way it does. Assuming that culture is just the product of text and context or of symbols and signs and ignoring the body and its history forced Sahlins to lose the explanatory potential to account for signs and symbols.

In turn, sociobiology reduces culture exclusively to biological mechanisms and neglects not only its social aspect that radically changes the scale of behavior, but the cultural one that exponentiates it by the accumulation of information in durable nonbiological, transgenerational formats such as books or monuments. Both lose, Sahlins and structuralism the material basis of signs and Wilson and sociobiology the acknowledgment of the emergence of qualitative new evolutionary orders from the critical mass of the social in culture.

Culture is built, we know. This book is a cultural artifact that remains in the ink of its pages or in screen bits and is transmitted to the reader because you are now building it word by word as I write it key by key. The front wall in front of you, of me, was also built brick by brick, and the city wall by wall with every bucket of mortar, sand, and gravel. As the kilo and a half of stones placed by the black wheatear in his nest, the kilo and a half of cells which is my brain is active now to give coherence and intelligibility to these ideas. It implies physical work connecting one perception with another, focusing our eyes, training memory, sharpening the mind to receive faithfully the ideas of other authors and organize sentences and their meaning as accurately as possible, always begging the mind to concentrate.

As I have mentioned, we cannot contrast evolutionism to constructionism because the latter continues the former at another historical and material level. As argued by Gould and evidenced by differentiated species, there are jumps and discontinuities in the process of evolution. Symbiogenesis of eukaryotic cells and the evolution of sexual organisms were jumps in evolution that, through the exhibition of phenotypes, involved selective or preferential acts. Also, the emergence of cumulative language that can be stored in material formats by a single animal species marked another evolutionary leap which constructionist theory (as it developed from pragmatism) can account for. Indeed culture is not autonomous or totally malleable because it emerges from evolutionary processes determined by morphology. It is not an automatic result of biology either, as it so often forces biology against itself. Sitting for 10–14 hours on a desk or at a computer, as one does nowadays, is a battle against our natural tendency to walk and change places.

Dobzhansky noted that cultural changes are too rapid to be correlated to genetic changes and states that "the premise which cannot be stressed too often is that what the heredity determines are not fixed characters or traits but developmental processes. The path which any developmental process takes is, in principle, modifiable both by genetic and environmental variables." He insists that inheritance is not a goal but only conditioning and certain genetic conditions can be desirable in an environment and not in another.

> What is more, since the environment in which man lives is in the first place his sociocultural environment, the genetic changes induced by culture must affect man's fitness for culture, and hence may affect culture. The process thus becomes self-sustaining. Biological changes increase the fitness for, and the dependence of their carriers on culture, and therefore stimulate cultural developments; cultural developments in turn instigate further genetic changes. This amounts to a positive feedback relationship between the cultural and the biological evolutions. (Dobzhansky 1963, 146–147)

More interesting than the debates between evolutionists and anti-evolutionists is the intra-evolutionists polemic. Lynn Margulis (1996) masterfully performed her move when she pointed to the fact that such alleged "new synthesis" is based on an insoluble contradiction: while evolution depends on change and variation, genes are fixed (that is the meaning "gene" maintaining its fixed configuration). This important point has not been, as far as I know, properly answered. We could even say that as light speed is the paradigmatic referent constant to physics, so are genes the constant to biology. In the Mendelian approach, organisms do not change over time, only their features recombine.

From Margulis's side of the field, not everything is random variation and selective retention. There is also variation by association or symbiogenesis. "My major thrust is how different bacteria form consortia that, under ecological pressures, associate and undergo metabolic and genetic change such that their tightly integrated communities result in individuality at a more complex level of organization. The case in point is the origin of nucleated cells (protoctists, animal, fungal and plant) cells from bacteria."

> While Gould and the others tend to believe that species only diverge from one another, I claim that—more important in generation of variation—species form new composite entities by fusion and merger. Symbiogenesis is an extremely important mechanism of evolution. Symbiogenesis analysis impacts on developmental biology, on taxonomy and systematics, and on cell biology; it hits some thirty subfields of biology, and even geology. Symbiogenesis has many implications, which is part of the reason it is controversial. Most people don't like to hear that what they have been doing all these years is barking up the wrong tree. (Margulis 1996, 136)

Apart from the fact that a proposal that considers evolution as a product of cooperation seems more kind and agreeable than the view of struggle for the survival of the fittest, empirical evidence has been accumulating that different type of genes exist within a cell. The strongest evidence for Margulis's thesis is the fact that mitochondria have a different DNA from the nucleus. Maynard Smith and Szathmáry question the romantic theory of cooperation and symbiosis of microcellular organisms and propose instead the cynical hypothesis of exploitation of the mitochondria in a cell as if kidnapped for the service of nuclear DNA. Would this imply that we have a case of mitochondria undergoing the Stockholm syndrome?

To take sides in the controversy between these various evolutionist approaches, the only logical move for a nonspecialist is perhaps paying attention to the octopus, as we will at the end of chapter 3, or better even, consulting the oracle of the octopus Paul who correctly predicted the outcome of the 2010 World Cup. While the outcome of this game is being decided, what I can truly attest without hesitation is that subjectivity, the interior view to the world, is the product of evolution and so can be found in simple as well as in complex creatures. Any living creature has a degree of subjectivity since autopoiesis depends on distinguishing self from nonself. Hence, more complex levels of feeling or experience in living creatures are expressed through the evolution of somatic, acoustic, and scopic registers to language in the human species.

TOUCH AND THE EVOLUTION OF THE SENSES

Better even than touching the sky with your hands is being touched in your hands by the sky. So expresses Michelangelo in the Sistine Chapel where by touch God bestows soul to man. In touch the origin of sensitivity and life reside, as aesthesis is primarily touch, skin con-tact (common touch). We are membraned beings, protected and at the same time exposed to the world by our membranes. We begin to know the world through touch. Our skin insulates us from others but also unites us with them in the most intimate manner, as if we would touch skin to skin. We live under the skin and right there we urgently feel the pungent need for others. The skin caresses, squeezes, beats, cools, calms, pinches, scratches, licks, comforts, hugs, kicks, bites, sucks, penetrates and receives, swells, hardens, contracts, kisses, burns, gets wet, freezes, seizes, itches, scratches, sweats, pales, blushes, gets diseases, is cured or poisoned, becomes velvety or dry, irritated or rots. The skin both protects and betrays us.

Ashley Montagu cites twenty biological functions of the skin, which are both semiosic and aesthesic. In the skin, the basics of sensory reception reside: information processes, barriers between the organism and its environment, mediators of sensation, and immunological protection of hormones and cell differentiation are all there. Skin protects injured parts against radiation and toxic materials, regulates blood pressure, is regenerative, produces keratin, is a temperature controller, and is involved in metabolism and fat storage. The skin is the largest sensory organ of the body, and every hour more than a million skin cells are replaced. (Montagu 1978, 6–9)

For Aristotle in Περὶ αἰσθήσεως καὶ αἰσθητῶν (On sense and the sensible) sensibility and the rest of the senses originate in touch. Touch is the most ancient, the most primordial, and perhaps the most sensitive of all senses considering a microscopic scale. How many steps in the evolutionary chain made possible the configuration of an entity that perceives the presence of another by contact? Robert Pollack argues for the perception even at molecular level through touch: "DNA too has words, syntax and meanings, but as a text it is a molecular LEGO set, a sculpture whose information—encoded in its very shape—cannot be read at a distance. It has to be 'felt' by other molecules in the cell in order to be read. DNA's letters are three-dimensional; they are read and understood by *touch*, the way a blind person reads a Braille text." (Pollack 1994, 22 emphasis added)

How and when was an impact occurring in an electromagnetic field translated as the "presence of another"? The simplest single-celled creature such as the paramecium is able to distinguish by touch what is edible from the inedible and signify it through its membrane. That starts the pendulum aesthesis-semiosis process that will cover the entire spectrum of evolution from the collision among molecules to their impact in the eardrum to the composition of symphonies.

Aristotle (*De Anima* Book II sect. 3) recognizes in touch the origin of the other senses when he writes:

> [A]ll animals have one sense at least, touch, and whatever has a sense has the capacity for pleasure and pain and therefore has pleasant and painful objects present to it, and wherever these are present, there is desire, for desire is just appetition of what is pleasant. Further, all animals have the sense for food (for touch is the sense for food); the food of all living things consists of what is dry, moist, hot, cold, and these are the qualities apprehended by touch; all other sensible qualities are apprehended by touch only indirectly. Sounds, colors, and odors contribute nothing to nutriment; flavors fall within the field of tangible qualities.

I say that touch is the very origin of aesthesis because it is the most archaic and primeval of all senses not only by the body's ontogeny in the fertilized cell but in the evolutionary and phylogenetic process (in the cell membrane receptors related to the ligands). Charles Darwin's grandfather, Erasmus, writes:

> [T]he first ideas that we become acquainted with, are those of the sense of touch; for the foetus must experience some varieties of agitation, and exert some muscular action, in the womb; and may with great probability be supposed thus to gain some ideas of its own figure, of that of the uterus, and of the tenacity of the fluid that surrounds it. [12]

> By 6 weeks the embryo, still lacking ears or eyes yet can perceive and touch on its lip and nose, and slightly move away from the source of stimulus; by 9 weeks it bends its fingers to grasp when touched in the palm of the hand. (Montagu 1978, 4)

From fish to humans, the mouth is the earliest part of the body sensitive to skin stimuli. Thus, we see that already in the womb, the fetus holds a finger in his mouth 5 months after conception. Lips and mouth are the most sensitive primary contacts before and after birth. He calls the "rooting reflex" the reaction of an infant of turning her head and mouth when the cheek or mouth are touched, which is a reflection of rooting and oral orientation, and it occurs also in response to the smell of the mother's breast. (Montagu 1978, 115, 121–123) This rooting behavior is consummated the moment when the lips meet the nipple and areola and absorb it. [13] We could perhaps understand this odd act of kissing that reminds of nourishing that probably stems from our cellular condition of a molecule incorporated by the membrane's receptors to feed. We kiss our beloved as if we would yearn to drink him.

In addition to the biological sense, there is a cultural sense of touch. Some people lack "tact" or sensitivity towards others' feelings, some things give us "goose bumps" and bristle hairs, we feel touched by something that moves us. Some situations touch the heart, we lose contact with reality. In the dictionary of the Real Academy of Spanish Language (RAE), there are 28 meanings of the word "touch." The skin is the body's external nervous system. Pain is a sensation of touch, a headache feels like someone squeezing or stinging us. Our skin protects us from harm, warns us; what hurts or soothes us is contact.

The stream of evidence that Montagu finds about the vital need for physical contact and the consequences of its deficiency is truly overwhelming, not only in humans but in different animal species such as rats, cats, and monkeys. Even armadillos, despite the difficult access to their skin, in order to feel cutaneous stimuli make a symbiosis pact with certain insects that enter into their shell and touch them. An animal that has not been touched is much more susceptible to disease, violence, and even premature death, because of a less effective immune system.

Contact thus alters even our chemical condition. Dogs and cats lick their newborns in the pineal region to activate their digestive and excretory processes. We have all witnessed the insatiable need of cats to be petted, especially those that have been rubbed by the mother. Even rats, when they have been touched, have shown greater emotional stability. (Montagu 1978, 239) The warm and intense communication among apes such as chimpanzees and macaques is just by touch through caressing and scratching each other generating beta-endorphin with a relaxing effect. The touch of a rough texture in the concavity of the cocoon is how a wasp mother indicates the direction out of it, as an error in exiting to the trunk of the tree can cause death.[14] Also in the case of aphids, the touch initiates their development into winged forms. Pierre Huber notes that after being separated by an interval of four months, some ants recognize others that had belonged to the same community and caress each other with their antennae. (Darwin 1882, 292) Isn't it amazing?

Through the skin and touch, Helen Keller (blind and deaf) could enter into a relationship with the human world by opening up the possibility of semiosis and language. By touching, she learned to trust, understand, translate, communicate, and live as a human being. The cognitive process goes from the aesthesis through the skin that little by little shapes "islands of consistency" that get stored in memory by a recognition pattern.[15] Such recurrence in perception can highlight the new and separate the emergent from the regular and predictable in the frightening initial world of chaos. The skin feels the most intense affection, rejection, or violence. Rape is atrocious not only because it rips the skin but the guts open, the inner skin. The urgency or disgust of touching and being touched is a very deep experience in humans and probably other species as well.

Language developed from the touched, the toucher, and the touching: the subject touches, the predicate is touched, and the verb is the act of touching. One touches with thought, with language, with sight. The body begins enclosed in the egg or the womb and opens up to orient itself, as a newborn to the nipple, contacting other open shapes.

Sight depends on touch as the photon reaches photoreceptor cells of the retina and is absorbed by invading their electromagnetic fields. Touching is about being in close proximity, penetrating another's territory; it is never contact at the atomic scale of matter with matter or particle with particle.

Seeing is literally a spectacular prodigy by the magic of evolution in which the photons are absorbed by the cell receptors of retinal rods and cones to generate neural signs in the cortical visual areas. The cellular receptor is preceptor and perceptor in contact with the photon and it acts as effector sending signals that interpret the impact. In a primate, more than 30 areas of the brain comprise the visual system, half of its brain. The evolution of the eye is a paradigmatic example for tracking the development of aesthesis: a series of molecules come together as peptides and proteins to form cells that differentiate themselves and can be organized in groups not only to detect photons, but also to perceive color, shape, distance, texture, and the rainbow. The events that make seeing possible have an impact at different scales: at the molecular scale the photon generates an electromagnetic field which excites proteins and mobilizes enzymes to release a cascade of neurotransmitters in synaptic connections across the neural web throughout the cortex. The tactility of clothing in Velazquez's paintings or van Gogh's brushwork expressivity shorten the distance between sight and touch and heighten the bonding with the artwork. Hearing is also tactile because the eardrum is excited by contact with molecules in the air, providing a cue to the environment and becoming a source of pleasure by melody, rhythm, volume, timbre, skill, and tone combined.

The primary purpose of perception is metabolism and safety, and it has evolved from bacterial membrane receptors that perceive solar energy, food, water, carbon dioxide, or electrons in surrounding molecules. Taste works through receptors on the tongue, palate, and epiglottis for distinguishing sweet as nutritious, from bitter which can be poisonous, sour which provides food and can aid digestion, and salty which contributes to assimilation. Recently a sixth sense called umami or tastiness has been added that detects the nutritional value of meat and fats. In other words, we detect molecules of sodium chloride (salt), carbohydrates (sugars), muriatic acid (acid), quinine (bitter), plus glutamate (umami) by means of which we signify stimuli as useful or harmful to be absorbed or rejected by the body.

The sense of smell also depends on touch since olfactory receptors recognize the physical and chemical conformations of aromatic molecules. This sense achieved more varied and finer distinctions than taste. It consists of a molecular coupling mechanism of olfactory receptor cells in the epithelium of the nasal cavity to the olfactory bulb of the forebrain. This mechanism operates to obtain information from the environment through the shape of aromatic molecules and produce intense emotional reactions such as disgust, longing, joy, euphoria.

The evolution of perception, both in phylogenesis and ontogenesis, has consisted in distancing ourselves corporeally from our antecedents which perceive by direct contact like the paramecium and the embryo, to those already capable of sight like a fly and a newborn, to the human species now able to see the embryo within the womb or the Big Bang 13.7 billion years ago.

BEAUTYING

> [A]nd again all the animals, a beautiful horse or cock or quail and all utensils and land vehicles, and on the sea freight-ships and ships of war; and all instruments in music and in the other arts, and, if you like, customs and laws also—pretty well all these we call beautiful in the same way looking at each of them—how it is created by nature, how it is formed, how it has been enacted—the useful we call beautiful, and beautiful in the way in which it is useful, and for the purpose for which it is useful, and at the time when it is useful; and that which is in all these aspects useless we say is ugly. Now is not this your opinion also, Hippias?
>
> —Plato *Hippias Major* (295d–e)

One of the firmest axioms of traditional aesthetics from Kant on, and in complete betrayal to Socrates's position is the antithesis of the beautiful and the useful. In the third moment of the *Critique of Judgment*, § XVII Kant concludes that "beauty is the form of finality in an object, so far as perceived in it apart from the representation of an end." He referred to the intrinsic value of beauty that does not, and should not, lead to any further purpose. Since Kant, this separation between usefulness and beauty has been jealously safeguarded by theoretical tradition to the extent of becoming a dogma. Such an immaculate conception of beauty seems to betray a longing for an idealized image of the bourgeoisie itself free from all worldly considerations and concealing the inelegant profit obsession of capitalism itself, as Eagleton (1990) acutely notes.[16]

This tension between the functionality and the gratuitousness of beauty resides at the very heart of the debate in aesthetic theory since Socrates playfully questioned Hippias on these issues. Hippias responded to the question on beauty as the most orthodox evolutionists might have predicted, because the example he chose to illustrate beauty is, typically, a maiden (as well as a mare, a vase, ivory, marble, or gold, which are all items useful for mating and status). Moreover, he also claims power to be the most beautiful and powerlessness the ugliest, concluding that whatever is more useful is more beautiful, for there is nothing more useful to Hippias than power. Plato *Hippias Major* (304 a–b)

Darwin, on the other hand, understood that beauty is useful but for a very different reason, namely that if it evolved it must have been for some purpose. Repeatedly he wrote that "females generally prefer or are more excited by brighter males; otherwise ornamented males would be of no purpose," which would contradict the evolutionary paradigm.

> Many will declare that it is utterly incredible that a female bird should be able to appreciate fine shading and exquisite patterns. It is undoubtedly a marvelous fact that she should possess this almost human degree of taste. He who thinks that he can safely gauge the discrimination and taste of the lower animals may deny that the female Argus pheasant can appreciate such refined beauty; but he will then be compelled to admit that the extraordinary attitudes assumed by the male during the act of courtship, by which the wonderful beauty of his plumage is fully displayed, are purposeless; and this is a conclusion which I for one will never admit. (Darwin 1882, 400)

> From the several foregoing facts it is impossible to admit that the brilliant colours of butterflies, and of some few moths, have commonly been acquired for the sake of protection. We have seen that their colours and elegant patterns are arranged and exhibited as if for display. Hence I am led to believe that the females prefer or are most excited by the more brilliant males; for on any other supposition the males would, as far as we can see, be ornamented to no purpose. (Darwin 1882, 316–17)

So we must distinguish which features are for display and which others happen to be attractive but were not meant to be exhibited: "Hardly any colour is finer than that of arterial blood; but there is no reason to suppose that the colour of the blood is in itself any advantage; and though it adds to the beauty of the maiden's cheek, no one will pretend that it has been acquired for this purpose." (Darwin 1882, 261) Darwin insists on the intricate relation between beauty and purposeness, as well as on the consciousness and intentionality of displaying such beauty.

> When we behold a male bird elaborately displaying his graceful plumes or splendid colours before the female, whilst other birds, not thus decorated, make no such display, it is impossible to doubt that she admires the beauty of her male partner. As women everywhere deck themselves with these plumes, the beauty of such ornaments cannot be disputed. (Darwin 1882, 92)

> Although so many pheasants and allied gallinaceous birds carefully display their plumage before the females, it is remarkable, as Mr. Bartlett informs me, that this is not the case with the dull-coloured Eared and Cheer pheasants

> *(Crossoptilon auritum* and *Phasianus wallichii);* so that these birds seem conscious that they have little beauty to display. (Darwin 1882, 400)

While Marx distinguishes between basic needs (food and safety) and secondary needs (artistic), Darwin distinguishes between primary characteristics for their direct purpose for reproduction (genitals, sexual organs) and secondary traits (song or birds' plumage, crickets stridulation, deer and elk's antlers). Note the parallel in the aestheticity of the second term, since all these are adaptations that contribute to seduce and distinguish one gender from another and one individual from another in sexual selection. Therefore, the evolutionary perspective would support Socrates's concept of kalokagathia καλοκαγαθία when he guides Hippias to recognize the convergent relationship between beauty and what applies to good use.

Let us suppose that beauty of ornaments is actually disputable in such cases as the male lizard's inflatable neck skin. What cannot be disputed is the utility of such features for attracting the female, which gives evidence to its purpose. Darwin concluded that for each appreciable object or feature there must be a subject appreciating it or a creature who appreciates it, unknowingly alongside Marx who argued that the production of objects or commodities implies also the production of subjects as consumers. A grouse's gill was an adaptation to attract the female because females must have been attracted to such a device if they gave progeny to gilled males (as the attraction to that gill is another adaptation that grouse mothers transmit to their daughters).

The evolutionary perspective sheds new, very concrete light on old problems. Following Hume's quest on the standard of taste beyond anthropocentrism, Darwin focused on specific natural, everyday behavior of each species and its attraction or repulsion to certain traits. He observed the choices that animals made to infer preferences and inferred the value that the chosen object could have for them.

Back to the Kantian position, in the second moment of his *Third Critique*, Kant concludes that beauty is what pleases universally without a concept. Here three additional factors intrude: conceptualization, pleasure, and universality. In Kantian aesthetics the "concept" is radically opposed to the sensibility, almost by definition setting the discussion on the ground of the nonrational. The second is pleasure—yet if the beautiful pleases, how can it not be useful as a provider of pleasure? The third term, universality of beauty, is set by Kant in the judgment of the subject from the s*ensus communis* or (as I understand it) sense in common of those who judge, not in the proportions and characteristics of the objects considered beautiful themselves. Yet, *communis* with whom? With a group of experts in the art world, with a social group, a species, or all live beings? Voltaire tends to favor the one before the last by stating:

> Ask a toad what beauty is, absolute beauty, the *to kalon*. He will answer that it is his female, with two large round eyes sticking out of her little head, a large and flat snout, a yellow belly, a brown back. Question a negro from Guinea; for him beauty is black oily skin, sunken eyes, a flat nose. Question the devil; he will tell you that the beautiful is a pair of horns, four claws and a tail. (Voltaire 1984, 63)

From this perspective, the basis for the universality of beauty is relative to the *sensus communis* of Guineans, frogs, and devils. [17]

As if it were not clear enough that beauty is an act of appreciation by a beholder and not an attribute of the object itself, consider a curious case that confirms Voltaire. Aborigines from the Malay Archipelago judge white teeth as ugly because they associate them with dog teeth and therefore paint their own teeth in different colors. Nor can women's thin lips be universally considered beautiful, as evidenced by the striking beauty convention among Mursi women of Ethiopia who deform their lower lips to hold a great disc. Young urban women today disfigure their lips with collagen to simulate female sexual peak fleshiness or pierce them with metal pieces that obstruct speech and eating. Beauty in a turkey's mucus, a grouse's inflated yellow esophagus, red bumps around some ducks' eyes, the reddened genitals of baboons and female chimpanzees, and the *Proboscis* monkey's (Nasalis larvatus) pendant nose, like collagen lips, is definitely in the eye of the beholder.

For Reid, the nineteenth century philosopher, beauty is very useful as an index of what men want in a woman: "What is beauty in the features of the face, particularly of the fair sex, which all men love and admire? I believe it consists chiefly of features that indicate good affections. Every indication of meekness, gentleness and benignity, is a beauty. On the contrary, every feature that indicates pride, passion, envy and malignity, is a deformity." (Reid 1801, 165) Reid's romanticism was a very unromantic calculation of each gesture as useful attributes of a perfect wife for a very timorous husband.

All these hypotheses and hints point to the idea that the sense of beauty could be none other than the very direction of evolution. I do not mean to imply that the purpose of evolution is to pursue beauty by some sort of aesthetic teleology. Beauty is not the cause but the consequence of evolution: what propels evolution, what we designate as beautiful marks a direction and furthers possibilities of thriving. It is the direction it points to that creates the sense of beauty. Evolution does not seek beauty as if it would be a pre-established value; it pursues itself and rewards itself with the pleasure of experiencing something we call "beauty" through preferences for symmetry, proportion, elegance, vitality, color, grace, dexterity. Beauty is related to vitality and life, as ugliness is related to sickness and death. The appreciation of beauty is an adaptation that allows us to distinguish what favors life and has the potential to convey it, and what obstructs it. The absolute value of ugliness is a decomposing corpse whose proportions have been altered; its smell fouled, its color dulled, its feel cold and hard in rigor mortis. Our sense of beauty marks the direction of life and evolution.

Steven Pinker sums up the idea of beauty as a reward button for contributing to evolution. However, beauty and pleasure are not identical issues, since beauty can cause pain too, and ugliness may provoke pleasure depending on the context (Goya's black paintings or the grotesque). Pinker (1997, 536) assumes a cognitive instrumental explanation of beauty in terms of "strong, clear, analyzable signals from interesting and potent objects" as do Orians and Heerwagen and Kaplan in landscape preferences for their information and shelter potential. (Orians and Heerwagen 1992; Kaplan 1992) Humphrey too elaborates this position stating that:

> [A]esthetic preferences stem from a predisposition among animals and men to seek out experiences through which they can *learn to classify* objects in the world about them. Beautiful "structures" in nature or in art, are those which facilitate the task of classification by presenting evidence of the "taxonomic" relations between things in a way which is informative and easy to grasp. (Humphrey 1973, 432)

To sum up, the main irrefutable argument that beauty is useful is the pleasure we derive from it. Yet, why does it provide pleasure? Because it is useful to evolution. *Pulchra sunt quae visa placen* (beauty is that which pleases) wrote Aquinas. If beauty is what pleases when perceived, do we simply call beauty whatever pleases us? Certainly not, as there is a value, besides pleasure, ascribed to it. As Socrates emphasizes in the epigraph above, beauty resides in the act of assigning value, almost a speech act: "*We say* that the entire body is beautiful . . . *we call* beautiful to . . . and *we affirm* that what is useful is beautiful . . . all that is useless *we call* ugly."[18] Socrates does not refer to objects themselves but to acts of assigning that category when we say and affirm that something is beautiful: it is thus an act of language, of languaging an experience.

Since beauty is also a matter of language we must then be as precise as possible. Borges (1968) writes in his story "Tlon, Uqbar, Orbis Tertius" that "For example: there is no word corresponding to the word 'moon,' but there is a verb which in English would be 'to moon' or 'to moonate.' 'The moon rose above the river' is *hlor u fang axaxaxas mlo*, or literally: 'upward behind the onstreaming it mooned.'"[19]

Similarly, what happens to the subject when he assigns this attribute or affirms that something is beautiful could be called "beautying." When seeing the splendor of what is felt as beautiful, hearing a melodious voice, or sensing a wonderful perfume we are "beautying" that experience and are beautied in that experience. This act is not equivalent to "beautifying" which means making beautiful an object that was not such, whereas "beautying" has nothing to do with the object itself as it denotes an event that occurs to the subject, an epiphany of "embeautyment."

Beautying expands our perception to the fullest so we may best appreciate something and find perfection through it. On those terms, assigning the attribute of beauty to something is an act of beautying, which does not change the object in any way except through our perception of it. The subject is bonded to what is present, enjoys it ("sensibility is joy" said Levinas) and beauties in it. Being moved by the splendor of a landscape, by a melodious voice, by elegance and dignity is an experience and an act of "beautying." Our sensibility is alive and active every day and in every act to perform, in Tlonian language, this uncommon act of beautying. Menelaus was about to kill his wife who caused so much calamity to the Trojans and Achaeans, but when Helena uncovered her breasts, Menelaus beautied in them.

To Seize Humanity by Its Wing

Discovering the exact location of pleasure in the brain means seizing humanity by its wing. It is the golden dream for marketing and politics, advertising, art, show business, sports, and the drug cartels that would sell their souls to the devil, if they had any, in order to acquire this power. It seemed as though we already had it in our hands when Olds and Milner performed the famous experiment that was supposed to have located the area of pleasure and reward in mice. Mice obsessively pushed the lever to stimulate with electrodes the septal and brain tegmentum areas as far up as the cingulated gyrus of the cortex until experiencing starvation and exhaustion before pressing the corresponding lever for food. There are also sites in the lower centers where the opposite occurs: animals do everything possible to avoid the stimulus, as well as neutral sites in which they do not try to get or avoid the stimulus. At the end, it was found that these mice were not stimulated at the pleasure center but rather at the demand center, and pressed the lever not for pleasure but for the urgency to do so. (Olds and Milner 1954; Levitin and Menon 2005)

Since the brain is organized according to specialized areas and particular neural maps, something of the sort of a "brain's pleasure system" has indeed been located in the nucleus accumbens that is linked to decision making generating dopamine neurotransmitters and the ventral tegmental area with connections to the hippocampus and the prefrontal cortex. Recent advances in neuroscience have shown the correlation of pleasurable effects on the secretion of opiates, dopamine, and oxytocin where dopamine produces desire and serotonin inhibits it.[20] These substances produce different effects in different combinations, but undoubtedly affect social anxiety, sexual desire, apathy, depression, panic, bipolar disorders, and euphoria.

Pinker points out that pleasure can be considered an evolutionary bonus that often occurs in relation to activities that benefit the organism's fitness such as obtaining food (bodily or mental) environmental information, evaluating prospective breeding partners, or scanning the right environment. For Symons "pleasure, like all experiences, is the product of brain mechanisms, and brain mechanisms are the products of evolution by . . . selection. The question is not whether this view of pleasure is *correct*—as there is no known or suspected scientific alternative . . . but whether it is useful."[21]

Against the emphasis on fitness by neo-Darwinism and sociobiology, Wolfgang Welsch (2004) proposes the topic of aesthetic pleasure from a perspective he considers Darwinian although it is more Wallacian. He writes that: "Only beings that are capable of pleasure can, at a higher level, also be capable of aesthetic pleasure. In this sense, hedonism is the basis for aesthetics. This consideration opens up a perspective and task for evolutionary aesthetics that goes beyond Darwin's approach." (Welsch 2004, online paper) In my view, it actually goes beneath Darwin's approach. If aesthetics is the culmination of hedonism, then it would have to account for other types of pleasure such as sexual, gastronomic, cognitive, ludic, oneiric, and even sadistic pleasure. He adds that: "I am suggesting a kind of pre-aesthetic analysis of the evolution of pleasure, confident that this might give us a better, genealogical understanding of the constitution of the aesthetic." Ironically, in his attack on neo-Darwinism, Welsch is a match to Pinker the staunch neo-Darwinist for this hedonistic conception of aesthetics. Personally I would rather opt for hedonic psychology pioneered by Daniel Kahneman and Amos Tversky if the goal is the understanding of pleasure.[22]

Yet pleasure is aroused not only by beauty, as we well know. Hatred is the source of much pleasure among individuals addicted to this kind of stimulant. According to Semir Zeki and John Paul Romaya (2008) the center of hatred is located in the medial frontal gyrus, right putamen, bilaterally in premotor cortex, the frontal pole and bilaterally in the medial insula and right insula, right premotor cortex and fronto-medial gyrus. They also point out that hatred and love, though with different patterns of activity, are shared by two areas: the putamen and insula. On this topic, Edmund Glasser (2009, 17) writes:

> I am left with the uncomfortable belief that much of hate springs from genetic factors and that, in the course of time, it is quite possible that the world will be afflicted with another individual whose talents for invoking hatred will be even greater than those of Hitler. I can only hope that this work will promote further thought and encourage a new point of view in hate studies and in opening new approaches to the understanding of the mechanisms of hate.

I too hope that the more we understand human sensibility and its evolutionary origins, the better we may cope with destructive elements that endanger the brief period of time we have to enjoy the gift of life. So in addition to genetic factors, we should also address the hedonic factors in the pleasure that can be observed as effect of hate exploited time and again by political and religious propaganda which seize a great part of humanity by its wing… perhaps a bat's.

Neuronal Aesthetic Societies and Their Frenzy for Art

"Beauty in things exists merely in the mind which contemplates them," writes David Hume.[23] For Symons "Beauty is in the adaptations of the beholder" but more accurately I think, for Semir Zeki it is in the neurons of the beholder.[24] Aesthetic issues necessarily point to neural studies as the most reasonable path for inquiry, because if anything is clear about aesthesis it is that it depends on our nervous system and its five senses. Since 1967 Zeki has been studying visual perception in primates such as rhesus monkeys and applied positron emission tomography (PET), functional magnetic resonance imaging (fMRI), and electroencephalograms (EEG) to study the visual brain in humans and other animals. Among his key findings are detection of functional specificity in certain areas of the visual cortex for color-coded cells and the presence of three cones in primates' retinas with special sensitivity to a particular area of the visual spectrum. (Zeki 1980) He claims that it is more pertinent to ask how a function is mapped than how the visual field is mapped, because perception does not work in analogic or iconic relationship with the perceived environment, but instead to an almost abstract digital scanning of relevant elements. Thus, the neuronal structure of visibility, due to the computations of brain functions, precedes the impact of sensory stimulation. Again, as in the calibration of the beautiful, the emphasis is placed on the subject rather than the object.

Zeki emphasizes that "knowledge belongs to the brain, not [to] the outside world." He manages to prove that aesthesis does not operate by a hierarchic mechanism where the nerve endings detect stimuli that would be processed and decoded by a cortical web but through perception and interpretation taking place simultaneously. (Zeki 2006, 251, 257) Note that this discovery empirically confirms my Möbius strip metaphor of the semiosis-aesthesis relation I mentioned in the Prologue. There are specific areas or modules in the brain that detect and distinguish particular objects such as edge or color, similarly to the finding of the module that recognizes human faces that is handicapped in patients with prosopagnosia, language areas of Wernicke and Broca that were located by studying patients with aphasia and the emotional prefrontal cortex areas discovered by Damasio (1994, chap. 1) in the case of Phineas Gage that seriously damaged his power of decision making. The better we'll know the molecular structure of nerve terminals and various types of neurons, the more we may be able to understand how we select stimuli, and how the physiology of cells in the visual field interprets information.

According to Zeki, artistic creativity and visual preferences in both production and reception involve the activation of the middle orbitofrontal cortex. (Zeki 2006, 262) Convinced of being close to locating the exact area that calibrates beauty, he argues that it can even be measured in degrees: the higher the beauty perceived the more active an area of the orbitofrontal cortex which regulates average preferences becomes.

Ramachandran and Hubbard (2003), followers of Zeki, propose to develop a neuro-aesthetics that would be better called "neuro-artistics," as they seek the universal forms to explain all art through ten principles:

1. peak shift or exaggeration,
2. isolation,
3. grouping,
4. contrast,
5. perceptual problem solving,
6. symmetry, abhorrence of coincidence,
7. generic perspective,
8. repetition, rhythm, order,
9. balance, and
10. metaphor.

Once again, as when we discussed Thornhill's decatomy in a previous section, we have an unexplained taxonomical criteria for proposing these and not other categories (as Borges's description of a Chinese encyclopedia). The peak shift, for example, refers to cases of exaggerated features such as cartoons; he illustrates it with the experiment of rats rewarded for detecting rectangles that react more strongly to stimuli that have a greater degree of angularity or differ more from the symmetry of the square. Examples presented on Boucher obese angels that exaggerate the feminine, are the result of the Mannerist style, but ignore everything else as in our bias to react tenderly to babies' features, to the roundness and smoothness of contours, and to the softness of pastel colors which are all biological adaptations. What seem exaggerated are Ramachandran's and Hubbard's ambitions and conclusions, because there are no formulas for art, even from an evolutionary perspective. An artwork is a social and perceptual framing in Goffman's (1986) sense of frames and keying for the production and circulation of certain objects or events qua art. Ramachandran's and Hubbard's work is based on a contemporary and highly conventionalized concept of art.

Nadal and associates argue that beauty is not localized on a specific area of the brain but is linked to a network distributed in the orbitofrontal cortex of reward, the dorsolateral and medial ventrolateral, cingulate, prefrontal, besides the hypothalamus and the limbic system that regulate our emotions.[25] Moreover, as Pert (1999 *passim*) argued eloquently, emotion and pleasure are not located at a single point but are strewn throughout the body via a peptidergic network.

When considering aesthetics as the science of taste, the logical method to follow would be surveys and comparative analysis of taste and preferences. Precisely for this purpose, Daniel Berlyne founded in 1965 the International Association for Empirical Aesthetics following an experimental psychology program initiated by Fechner (1871 *Vorschule der Aesthetiker*) and Wundt (1874 *Grundzüge del Physiologischen Psychologe*). The *Journal of Empirical Studies of the Arts* publishes the results of these experiments and surveys about preferences in landscapes, faces, and bodies in comparative studies and exchanges between different ethnic, sexual, and generational groups. Other authors such as Robert Solso, Colin Martindale and David Rindos, William Hirstein, and Christopher Tyler work a parallel or convergent empirical approach to studying preferences. Martindale and Rindos propose the notion of quasi-random variation and hedonic selection as a cultural mechanism, and they elaborate a list of twenty-five aesthetic effects from a neuronal perspective to find isomorphism between creativity and perception of beauty. (Martindale and Rindos 1986) Terrence Deacon assumes that the aesthetic faculty is superior to perception and emotion and involves something like an "aesthetic evaluation module" derived from Fauconnier's "blending theory." (Deacon 2006, 21–53) Merlin Donald, for example, categorizes art as cognitive engineering. He states that art is universal and has three cascaded stages: mimetic, mythic, and theoretical, although inexplicably Aristotle's theory of mimesis was not even considered. (Donald 2006)

Raymond Tallis (2008) ridicules neuro-aesthetics' claims in explaining artistic issues because, among other reasons, it is unable even to make the most basic distinctions, for example, between reading one poem by John Donne and another, between successive readings of a particular poem, or between Donne and other metaphysical poets, between reading great works of literature and literary trash, between reading and a large number of other activities, such as getting toilet paper.

These various neurological and empirical approaches coincide in their assumption that the holy grail of aesthetics might be located in a specialized module in the brain able to specifically detect beauty and art. They dream of finally tracking down the *Sense of Beauty* alluded to by Darwin (1882, 92): "I am referring only to the pleasure generated by certain colors, shapes and sounds, and can properly be called as the sense of beauty." To the extent that societies of neurologists communicate in as a friendly and cooperative manner as do neuron societies, we will advance the process of understanding these ancient enigmas and crack the code for the cerebrome. For now, we have no choice but to start in the beginning.

NOTES

1. Quoted in (Smith 1990, 446)
2. My gratefulness to Arnold Berleant for bringing this book to my attention in the nick of time.
3. I have argued this position at length in (Mandoki 1994; 2006; 2007)
4. Converging with this basic concept of aisthesis, Berleant (2010, 9) holds that "nothing is more primary in human experience than sensory perception, and the satisfactions and dissatisfactions of experience are a principal motivation in our behavior. I take this primacy, then, as the originating idea of the aesthetic, aisthesis, literally, perception by the senses. This merging of the aesthetic with the activities and objects of human life must be acknowledged in aesthetic theory as well as recognized in human practice, not merely as a theoretical claim but also as demonstrable in the elucidation of different situations."
5. In this sense, this work could seem to converge with Shusterman's somaesthetics, (Shusterman 2008) especially by the common pragmatist approach, except that for Shusterman it is a means for self awareness following East Asian practice of meditation and discipline, while this work pursues a radically different cognitive biosemiotic and evolutionary approach.
6. Again, I must refer the reader to (Mandoki 1994; 2006; 2007) where I extensively argued these points.
7. On incident and bullying against Wilson, cf. Laland and Brown (2002, 97).
8. On a more thorough discussion of Sperber's points on this issue cf. Mandoki (2002).
9. Part of this anti-subject group in line, for example, with Paul Churchland who cancels even the possibility of experience and suggest brain states only. (Churchland 1981)
10. See for example Trivers's (1985) and Wright's (1994) work on altruism, Jablonka's and Lamb's on animal culture (2005), and Wilson's (1980) compilation on group organization among different species.
11. On primates' social, political, and emotional world cf. Sheets-Johnstone (1994); de Waal (2007).
12. Quoted in Montagu (1978, 253).
13. See for example the concept of latching on, bonding, or *prendamiento* that I have developed previously as a corporeal phenomenology of aesthesis. (Mandoki 2006; 2007)
14. Cooper 1957 quoted by Wilson (1980, 96).
15. Escalona writes "with recurrence, there develop islands of consistency" quoted by Montagu (1978, 251).
16. This theoretical prejudice perseveres even in evolutionary perspectives a quarter of a millennium later, as in Brandt (2006) who opposes aesthetic perception oriented to contemplation and affects against pragmatic action-oriented perception.
17. Contiguous (or should I say "common") to this *sense in common* is Berleant's (2010, 209) idea of "perceptual commons" that he defines as "the most inclusive environmental condition of human life. By the simple fact of living we are embedded in a perceptual sphere, and it is from here that we must proceed in order to function in that world." He adds that "perceptual commons is not private nor is it public. It is common." I would add that such perceptual sphere is in fact embedded in our body as in other creatures' morphology. Moreover, it is a species' perceptual common as well as common between environment and body that perceptually evolved attuned to it. The body is embedded in the environment as the environment is embedded in the body's possibilities of perception.
18. See the epigraph mentioned earlier in Plato *Hippias Mayor* (295c–e).
19. Jorge Luis Borges (1968). Can be read online at http://art.yale.edu/file_columns/0000/0066/borges.pdf. Accessed June 16, 2014
20. On the opiates and pleasure system see Pert (1999).
21. The clearly pragmatic approach of Symons quoted by Thornhill (2003, 13).
22. See for example Kahneman and Tversky (1973) and Kahneman, Diener, and Schwarz (2003)
23. David Hume's Essays, Moral and Political, 1742 (Hume 1965).
24. Symons (1995), quoted by Thornhill (2003), 22.

25. Nadal et al. (2008). http://www.science-of-aesthetics.org/proceedings/IAEA08%20 Proceedings%20Thursday @30Aug21.pdf#page=9. Accessed June 16, 2014.

Chapter Two

Orbis Primus

בְּרֵאשִׁית

In the beginning, God created the heavens and the earth in one single day. On the second, She separated the waters. On the third, She created plants, and on the fourth, the sun and the moon. On the fifth, birds and sea creatures sprang forth, and on the sixth, earth animals—including humans—were shaped. The order of the sequence seems a bit surprising, since at first glance it appears to obey an evolutionary logic following Oparin by originating life from water and light in a quite reasonable progression, except for that third bizarre cosmological day of plants without sun. A more logical order would have begun with the creation of the heavens and light on the first day, the distinction of sun, earth, and moon on the second, the separation of waters on the third, sea plants on the fourth, birds and earth animals on the fifth, and humans on the sixth (which is more or less what science proposes). Perhaps God has continued working secretly on the eighth, ninth, and tenth until today, but since we are creatures of the sixth, we are unable to conceive Her subsequent creations, just as creatures of the fifth day may be unable to acknowledge the singularity of humans. Or do they discern human mental uniqueness?

There is another version so mystical that it makes Genesis look like a scientific textbook. According to this exegesis, in the beginning there was a point of Singularity with absolute density in an infinite gravitational field where a great hot explosion occurred that in a millionth of a millionth of a second catapulted photons and radiation. In this cosmological myth called "Big Bang" the Everything of infinite concentration occupied a single point out of time and space, and said "let there be world" and it burst out vomiting quarks and electrons at tremendous speeds creating space-time by their expansion and movement. At 10 to minus forty-three seconds, the temperature decreased and the quark and antiquark dominated; antiquarks then disappeared, three quarks managed to join by gluons to form protons and neutrons that formed atomic nuclei of hydrogen, helium, deuterium, and lithium. Three minutes after this singular epic, matter and radiation were coupled and so lasted this instant of 300,000 years. Matter decoupled from radiation and electrons rebelled to join nuclei and generate a huge variety of autonomous organizations registered in the Mendeleev periodic table (which now numbers 118, along with those man-made, at the time I checked Wikipedia June 23, 2009). In triple time, a million years after the Big Bang, matter gradually assembled to constitute stars and quasars, even proto-galaxies and solar systems. This narrative, however, does not tell us if before the Hot Big Bang there was a Big Crunch or a Small Cold Bing, but what is established and accepted in this myth is that less than one second after the Big Bang God said: "Let there be a photon" and God saw that it was good and it dominated the universe.[1]

In this almost delusional interpretation, the first thing that stands out is that instead of continuing to expand to infinity, matter begins to clump together in tiny orders of small differences by bringing two or three quarks to form hadrons, then several hadrons into atomic nucleus. Electrons are attracted, atoms organize one, two, or three up to tens of protons and neutrons that seem to defy entropy, the sacred law of physics, since they are progressively ordered and differentiated. The energy in turn acts by strong and weak forces into atoms and particles, as well as electromagnetism and gravity, each operating by photons or gravitons in different patterns. These four types of energy make all those who seek a common formula very uncomfortable.

If it were true that God does not play dice, as Einstein claimed, She would have managed to invent another game as She has since then become bored of playing only with quarks, electrons, and radiation. She played out of curiosity and combined them into neutrons, protons, and all the existing elements like a LEGO game. What need was there, in the expansion of the universe, to try to form combinations of sub particles and atoms which in turn are coupled into molecules? If that complexity did not suffice, the combinatory of matter reached an unexpected threshold where molecules organize in templates that copy themselves, namely, replicating molecules. Due to its stability, the carbon atom's emergence allows the formation of amino acids that are linked in chains of nucleotides and proteins to give rise to cells, tissues, and organisms. Such diversification of matter could also have been a purely random, godless play of dice. If that would actually be the case, then Dice should be deified and Einstein excommunicated.

I understand, dear reader, that you may be quite dizzy at this point if you followed me until here, wondering when we will finally land upon aesthetics. Actually, we were here from the beginning when exploring the history of a photon, since it is the condition for aesthesis and origin of all sensitivity and all senses, as they have evolved precisely for detecting photons. If God said, "let there be a photon" it must probably have been because She sensed it.

THE PERFECT TAXONOMY

Borges cites weirdest efforts to achieve a perfect taxonomy. He mentions the enigmatic Chinese taxonomy as well as the bewildering Bibliographic Institute of Brussels according to which, as quoted by Wilkins:

> The universe has been parceled into 1000 subdivisions, of which number 262 corresponds to the Pope, the 282 to the Roman Catholic Church, the 263 to the Lord's Day, the 268 to Sunday schools, 298 to Mormonism, and 294 to Brahmanism, Buddhism, Shintoism and Taoism. Heterogeneous subdivisions are not rejected, for instance, the 179: "Cruelty to animals. Protection of animals. Mourning and suicide from the point of view of morality. Various vices and defects. Virtues and various qualities."[2]

Criteria for organizing a taxonomy may vary, but the impulse to order the world in a regular and reasonable sequence, hopefully proportionate and as symmetrical as possible, whatever its logic, overwhelms us: it may proceed from the simple to the complex, from the precarious to the abundant, from fixed to dynamic, from ancient to modern, bottom up, good to bad, tiny to huge, left to right, one p-brane to eleven dimensions that could have been twelve as the months or thirteen as the Mesoamerican calendars and the Metatron with its kabalistic value. Zerubabel (1993 *passim*) reveals the curious scruples we display to set totally arbitrary limits and not infringing them.

Each generation learns to categorize the world in a particular way. For the ancient Aztecs there existed four worlds before this, each dominated by different creatures: monkeys, fish, birds, and giants (in that order). Our elementary school teachers taught us about three kingdoms: mineral, vegetable, and animal. Our kids learned the Whittaker classification into five: monera, protista, fungi, plantae, and animalia. Decade and a half later, this taxonomy added to monera bacteria and archaea as well as eukaria.[3] The most recent and cryptic classification is Fabien Burki's 2007 into excavata (free and parasitic protists), opistoconta (amoebas, fungi, and animals), plants (green and red algae and plants), and SAR (acronym for stramenopiles, chromalveolates, and rizaria). (Burki and Shalchian-Tabrizi, 2007) Another recent classification seems much simpler, a mere dichotomy of eukaria bikonta (apusozoa, archaeplastida, chromalveolates, rizari, and excavata) and unikonta (amoebozoa and opistonkonta with metazoa, choanozoa, and eumycota), based on the molecular organization of cladograms according to the common ancestor. (Brusca and Brusca, 2005)

And since we are already engaged in taxonomic imagination, I am quoting Foucault's quotation in the Preface of *Les motes et les Choses* who quotes Borges quoting Wilkins and Dr. Franz Kuhn who quotes that famous Chinese encyclopedia entitled Celestial Emporium of Benevolent Knowledge and its bizarre classification of animals: those belonging to the Emperor, embalmed, trained, suckling pigs, mermaids, fabulous, stray dogs, those included in this classification, those that agitate as crazy, the innumerable, those drawn with a brush of very fine camel's hair, those which have just broken the water pitcher, those that look like flies from afar.[4]

Why baffle the reader with such exotic taxonomies? It is just a malicious trick to contrast them to a much simpler taxonomy required to avoid an order of reality that gobbles up one another with all their delicate and precise differences. The taxonomy I am proposing here has only three sets, and not for a trichotomania of sorts that several authors have evinced (typically Plato, Hegel, Charles S. Peirce) but out of the great lesson that neo-Darwinism and sociobiology involuntarily left by neglecting thresholds and discontinuities between the inorganic, organic, and cultural spheres and reducing it all to the dynamics of gene replication.

We will designate the first of the worlds, the physical universe of matter-energy and space-time by the plain term *Orbis primus* or first world, literally.[5] What Elohim labored upon the first, second, and fourth days of Genesis, and what pre-Socratic philosophers like Thales, Empedocles, Anaximenes, and Anaximander pondered about, as well as their successors Galileo, Newton, Einstein, and other Olympian gods of physics is this *Orbis Primus*. Around 13.8 billion years ago *Orbis primus* emerged, according to the last recorded data in Wikipedia (imagine what Borges would have done with the conjunction of a mirror as Internet and an encyclopedia in time as Wikipedia!).

To denote infinitely smaller areas nested within *Orbis primus* encompassing the third, fifth, and sixth days, according to Genesis, we will use the term *Orbis secundus*. This second world embraces life as was cooked by light in the primordial soup of the oceans 3.7 billion years ago according to Oparin-Haldane. Biotic communities of bacteria and archaea evolved from prokaryotic to eukaryotic comprising all multicellular organisms about 1,400 million years ago.

And since we are going from infinite to infinite, in certain infinitely smaller areas within the *secundus* a third world emerged by the accumulation of mutations, this time intentional, of inherited transmission: *Orbis tertius*. As genetics in the *secundus*, it also depends on messengers, codes, and interpretations, but it can skip several generations, die completely or remain piecemeal, recessive or dominant to reappear centuries later. The *tertius* is cultivated by cults around symbols such as deities, heroes, forefathers, gold, or power. It is a collective phenomenon (as multicellular organisms are collective, and as DNA is a collectivity of nucleotides) that is constructed according to various scales, from a couple to the family, tribe, clan, neighborhood, gang, sect, town, city, state, country, empire, and civilization. What is common to *secundus* and *tertius* is that both emerge, inhabit, are fed from and depend on the *primus*.

As one cannot explain biology by the laws of physics alone, or culture only by the laws of biology, we cannot ignore them either. This is not to plea that each world has irreducible essences but rather that particular configurations of matter and energy are at work in each case. There seems to be a common dynamic to these three orbs: a process of increasing diversification and complexity where quantitative changes turn into qualitative changes of a different order (as Marx observed in historical social change). Particular molecular couplings permit the emergence of unicellular organisms some of which generate assemblages of cells, which in turn enable multicellular individuals and later multi-individual organisms by symbiogenesis that can consecutively establish colonies or populations. Such qualitative changes accumulate over time with a clear sense of directionality because we may find molecular links in the cell, but not cellular links in molecules, nor cultural links in cells or organisms. This directionality requires to be examined as a specific issue of thresholds, filters, and irreversibility. Hence, we must consider a temporal and a structurally progressive order from *primus* to *secundus* to *tertius*.

The borders between these three orbs are porous and blurred, since liminal creatures inhabit between the *primus* and *secundus*, hybrid things/beings such as prions that are misfolded replicating proteins (causing lethal effects such as the mad cow syndrome), or fungi yeast and replicating viruses with genes and proteins. Ribosomes, which are molecular manufacturers of protein, appear to "choose" in the replication and methylation processes which genes to activate and which ones not, could earn an emigration visa from chemistry to biology. Bacteria communicate by local and bacterial codes since semiosis begins in cells and continues through various multicellular organisms to species' social groups up to culture.

The threshold between *secundus* and *tertius* is exemplified by Imo, the female chimpanzee inventor (or inventress) of washing potatoes and wheat, teaching her discovery to the other members (especially females) in her community. (Kawai 1965) In the three orbs different kinds of disasters can strike: imploding stars and volcano eruptions in the *primus*, epidemics and plagues in *secundus*, and wars and genocides in the *tertius*.

The *primus* is moving. The *secundus* is in activity because it deals with the labor of all creatures in their survival. The *tertius* is dedicated to the work which is cumulative and bequeathed to the next generation.[6] Broadly speaking, this is the rose of the three winds with which we can orient ourselves along this journey, even if it designates incommensurable scales (except in death, which returns us all to *Orbis primus*).

SPHEROLATRY AND ALL ITS DEVOTEES

> Nature is an infinite sphere whose center
> is everywhere and circumference nowhere.
> —Pascal

Xenophanes of Colophon, a prophet sadly ignored, proposed a spherical god monotheism (sadly because we are all somehow spherolaters, cryptic or confessed, and have not credited him). Plato saw in the sphere the most successful of geometric shapes that encompasses all others, the image of perfection and uniformity. He was not original in this, as for Parmenides the sphere is self-sufficient and defines the entire world of Being since a sphere is accurately fixed at each point, identical to itself, perfect, never larger or smaller in any of its parts. For Empedocles, "the round sphere that enjoys in its surrounding solitude is the perfect stage of Being, in which friendship dominates and is the same everywhere."[7]

Also in his view of the universe, Pythagoras imagined the earth as a sphere located at the center of the universe surrounded by the heavenly bodies, each attached to a glass sphere whose rotation produced wonderful musical harmonies. We have become deaf to the harmony of spheres by getting used to it from birth, while only Pythagoras was able to hear it, unfortunately without the code to register it in scores. What we are now able to hear is the monumental organ of the solar atmosphere.[8] Astronomers Taroyan and von Fay-Siebenburgen have discovered that the sun plays heavenly music of acoustic magnetic waves similar to a pipe organ. Indeed, that music "has been recorded and reveals harmonious sounds caused by the movement of giant magnetic loops in the solar corona—the mysterious and least known outer layer of the solar atmosphere. More importantly, the team studied how the sound decays, providing an unprecedented insight into the physics of the solar corona."[9]

In the atomic model of Leucippus and his pupil Democritus prevailed a spherical image of the universe. Aristotle wrote that "of all shapes, the spherical shape of the atoms is the most mobile, and therefore the figure of the particles of fire and mind." "The soul is a kind of fire or hot substance, its 'forms' or atoms are infinite in number, those which are spherical he calls fire and soul . . . the spherical atoms are identified with soul because atoms of that figure are the best adapted to permeate everywhere and setting everyone else in motion by being themselves in movement." (Aristotle *De anima* II 3 and I, 2).

The incorruptible crystalline ether spheres surrounding the world described by Aristotle dominated the picture of the universe until the Renaissance. Hipparchus proposed a model of celestial bodies moving in circular orbits around the spherical earth. This model later influenced Ptolemy's geocentric system with seven circles of planets. The spherical model of the earth by Al Idrisi in the twelfth century and Johannes Sacrobosco's *Tractatus de Sphaera* in the fifteenth century are other versions of this human passion for spheres. It's not surprising that in his heliocentric model, Copernicus had less scruples in abandoning the centrality of the earth than in giving up the circular shape of planetary orbits. It took Johannes Kepler who, despite an aesthetic sense of spherical perfection, very reluctantly abandoned the hypnotic charm of the circle as the mathematical and empirical evidence of calculations and observations mounted, forcing him to propose an elliptical model of the planetary system.

As late as the twentieth century, the "plum pudding" model by Joseph John Thompson, Nobel Prize for discovering the electron, was visualized as a spherical shape and embedded in another area of positive charge. Bohr's atomic model remains true to the field of particles with electrons spinning circularly around the nucleus. Alexander Friedmann proposed that the universe looks the same from any direction from which we observe. He recaptures the spherical model of Parmenides, although unlike that static model the Friedmann universe is expanding like a balloon that is inflated. Werner Israel proposed in 1967 that all black holes are spherical and in 1986 Michael Duff developed a model of almost spherical bubbles that float in 11 dimensions.[10]

Quarks and antiquarks dethrone the proton and neutron as elementary particles, yet the spherical shape representing those still remains. The sphere is the simplest figure for describing orbits around the nucleus, and the field in which wave-particle moves for Schrödinger's quantum mechanical model of the atom. Protobiotic fatty acid vesicles in Szostak's model are spherical, and so are the elements in the flowers and stars of Shechtman's quasicrystals as well as Riemann's sphere and the orbital probability for finding an electron.

Such is our aesthetic fascination with the sphere that the Goseck Observatory still retains a perfect circle of 75 m. diameter built 7000 years ago to watch the movements of the stars.[11] The sphere dominates our perception of the moon and the sun, the ovule and fetal life encompassed in a quasi-spherical uterus and the human figure illustrated by Leonardo da Vinci. If we are set to the divination of the future, what is the most logical instrument to do it? A crystal ball of course! Sufi dervishes attempt to form a sphere in their dance swirling around. The mystical *sephirot* concentrate the sense of divinity for kabbalists. Spherical balls are the universal childhood game. The sacred ball game practiced in several Mesoamerican cultures, the circular so-called Aztec Calendar, the Mandalas of Hinduism, the rosettes of Gothic cathedrals in Notre Dame and Chartres, and the interior of the Pantheon in Rome show the extent to which the spherical and circular model has dominated our imagination.

The first representation of the human figure by children is usually a circle with two points, the eyes. We are captivated by soap bubbles; balls hypnotize us in a game and we have invented all sorts of rules for playing with spheres: marbles, bowling, billiards, bocce, golf, baseball, tennis, or football. It is a tradition to hang colored spheres from cut pine trees during the winter to celebrate pagan and Christian festivities. The circle and the sphere are also favorite logo figures of Mozilla Firefox, Google Chrome, Wikipedia, Merriam-Webster, Mercedes Benz, Heineken, VW, Ubuntu, Starbucks, the Olympics, and Alfa Romeo.

Spherical shapes provide maximum volume with minimal surface. Thus in evolution plants produce spherical fruits to optimize size and protection from animals before they mature. Perhaps the root of this fascination could be not only in the vision of sun and moon but also perhaps in the mother's nipple. Currently spheres continue to delight science and philosophy as metaphors bringing forth spherical categories as the exosphere, troposphere, thermosphere, mesosphere, stratosphere that make up the atmosphere. Eduard Suess proposed in 1875 the tripartite division of atmosphere, hydrosphere and lithosphere and Vladimir Vernadskiĭ coined the terms of geosphere, biosphere, and noösphere which Édouard Le Roy would later develop as the field of transhuman thoughts or mental interactions on earth and the effects of mind over matter. (Vernadskiĭ and McMenamin 1998) With the concept of *umwelt* as a bubble, Jakob von Uexküll (1982) denotes the perceptual world that surrounds each creature and determines its ability to perceive and act. Umberto Eco points to the contemporary world as an iconosphere (Eco 2004; 1976, 367) and Yuri Lotman (2005, 5–23) coined the concept of "semiosphere" as a model for the analysis of culture. So the sphere of influence of the sphere is immense.

In his evocative title "Music of the Spheres," Thomas Sebeok (2000) meditates on the origin of the concept of "semiosphere" to explain the semiotic relations on a particular location. The attraction of the spherical model was still present in his choice for the title "The semiotic sphere."[12] A year later, Sebeok briefly mentions the decision to change the metaphor to that of a web (semiotic web) of semioticians.[13] Such quasi-Keplerian surrender was fortunate for many reasons and greater consequences than those he probably foresaw. It may not have required more than the flutter of a butterfly on Sebeok's desk for the semiosphere to be replaced by the semioweb paradigm of semiosic processes. But since then, this small trifling particular entails enormous consequences for semiotic theory.

The accuracy of an explanatory metaphor is of great importance for the development of a theory, although, as noted by Norbert Wiener "the price of metaphor is eternal vigilance" because we can easily take them literally.[14] For George Lakoff and Mark Johnson (1980) metaphoric mappings allow clarifying a more abstract concept like target by means of a better-known concrete source on the basis of a body experience. As these projections have degrees of systematicity that impose limits on the relationships between the domains in which they apply, we must follow Wiener's advice and watch them almost as much as our lies to avoid inconsistencies. Spherical metaphors are applied at different scales: Lotman's semiosphere is social and cultural, while von Uexkül's *umwelt* is individual and biological or a subjective reminiscent of Leibniz's Monadology except that the *umwelt* itself depends on the morphology of the organism and lacks monads' uniformity.

We are consequently faced with many questions: How or in what manner does an *umwelt* relate to a semiosphere? To Kalevi Kull (1998) a semiosphere "is the set of all interconnected *umwelten*. Any two *umwelten*, to communicate, are part of the same semiosphere." Of what nature are their interactions? While the concept of *umwelt* revolves around a subject, it is not clear where its place in the semiosphere is. As Pythagoras's universe of concentric crystalline spheres, the model fails to explain what remains outside semiospheres. Aristotle's celestial spheres lasted almost two millennia until Galilean and Newtonian physics broke them to pieces. But the sphere metaphor is alive and well and fascinating us as spherolatrous creatures that we are, hence the origin of the *orbii* (circles) *primus*, *secundus*, and *tertius*.

WEBS, RHIZOMES, AND THIN SHELLS

The revelation that the bulk of green foliage on a tree is actually a web of branches, tiny ribbed twigs and leaves, mosses, ladybugs, spider webs, caterpillar cocoons, bird nests, butterflies, robins, lichens, and dew drops all intricately entangled has been an experience common to many short-sighted people when we see the world for the first time with our glasses on. Spheres turn out to be a rich configuration and interaction of elements at other scales and dimensions as a Buckminster Fuller's geodesic ball of webs, or architectural structures with thin skins such as Ingalls Rink and the Trans World Flight Center by Eero Saarinen, Nagoya dome, or the roof of Two Sisters velodrome. Equally, that huge bright ball hanging at daytime from the sky is also made of webs in frantic activity, atoms of hydrogen and helium, oxygen, carbon, iron, sulfur, neon, nitrogen, silicon, and magnesium interacting and playing this monumental organ by loops in the corona of that magnificent celestial body known as "the sun."

The theory of matter took the task of seeking minimal and pure isolated units of elements represented as spheres from Leucippus and Democritus's atoms to Murray Gell-Mann and George Zweig's quarks of different flavors and colors. This minimalist view contrasts with the mythological vision where everything appears to be entangled: cardinal points, colors, stones, months, worlds, temperaments, and the Aztec gods; men and totemic animals, stars and human horoscopes. In contrast to the atomistic view of the individual as isolated sphere hopelessly rebounding against society which does not comprehend his profound and unique sensibility (in romanticism's version), the web has gained ground as the dominant metaphor for our society (since neural webs' models and Arpanet, intranet, Internet, World wide web developed) in which an individual is located at nodes across different scales, rather than opposed to the collective. This reticular paradigm was also behind Deleuze and Guattari's (1966) suggestive essay on rhizome that toppled the exclusivity of the hierarchical tree as dominant metaphor for various explanations in humanistic studies. In this context, the concept of "semiosphere" appears like a photogram or single frame isolated for analysis in the dynamic flow of the complex semiosic network. It is the green ball of foliage that keeps inside worlds and underworlds.

The web model has also affected genetics, because each gene, as a minimal unit, was supposed to be responsible for certain effects. The decoding of the human genome in 2001 showed that, on the contrary, such effects depend on webs of relationships between genes and their environment at various scales, because the genome has a systemic structure. (Jablonka and Lamb 2005, 7) In other words, genes are context dependent and consequently not isolated units but inmeshed in a genetic web that includes its environment.

Half a century earlier Ludwig von Bertalanffy (1969), the initiator of the general systems theory or GST, considered properties at various levels as part of a comprehensive system. The semiosphere model is also a dynamic flow within and across a semio-web, so returning to the problem on what would remain outside semiospheres, we can find an answer if we drop the closed spherical container metaphor. The semio-web then is an open, dynamic, evolving, emergent, interconnected, heterogeneous, and changing system that, as the universe, creates space-time by its own expanding motion. Since webs are decentered semio structures, the concept of center and periphery in Lotman's semiospheres has to be relativized to symbolic nodes of attraction or repulsion at various strategic points of semiosic space/time. Thus, a web arises and grows asymmetrically from any of its parts, as a rhizome, in contrast to the sphere that must be blown symmetrically from a single central point. As Borges writes in *The Garden of Forking Paths*, "He believed in infinite series of times, in a growing, vertiginous web of divergent, convergent and parallel times. This web of times which approach one another, fork, cut off or that are secularly ignored, embraces all possibilities."[15]

In the *orbis primus*, particles can be explained as a result of vibrations of different tones as proposed by the superstring theory. Also, the mere fact that carbon atoms have four electrons in their last orbital level enables them to keep relatively stable chains and create webs with other atoms through very complex organic compounds. In *Orbis secundus* these carbon chains allow replicating nucleotides forming peptides and proteins to generate cells, tissues, organs, and systems such as the circulatory, immunological, endocrine as well as a trillion synaptic connections associated to other people's neural connections through cultural semiotic systems like interwoven language and visual codes exponentially projected through time by material documents. All life activity is co-activity and connectivity linking contextualized interactions with other creatures, even with their past and future evolution. Adaptation processes are always two-way co-adaptation of organisms to their environment that consists of other organisms coadapting among themselves.

> Both biological and technological evolutions consist in the invention of slightly or profoundly novel organisms, goods or services which integrate into the ecological or economic mesh and thereby transform it. Yet at almost all stages, the web retains a functional coherence. Furthermore, the very structure and connections among the entities set the stage of web transformation. . . . The web governs its own potential for transformation. (Kauffman 1993, 370)

> The vision of living systems and webs provides a new perspective on the so-called hierarchies of nature. Since living systems are webs at all levels, we must visualize the web of life as living systems (webs) interacting in a web with other systems (webs). For example, an ecosystem can be schematically represented as a web with a few nodes. Each node represents an expanded body and appears as another web. Each node in the web represents an organ, which in turn will appear as a web to be amplified and so on. (Capra 2002, 54, 58)[16]

Webs characteristically consist of relationships among parts rather than of objects, a dynamic rather than the individuality of its constituents, and consequently requiring a qualitative rather than a quantitative description. In a web, the relationship of scales becomes a priority criterion, as the recurrence of similarities at different scales defines their nature (i.e. the Julia and Mandelbrot fractals or pentadic quasicrystals in Shechtman). Webs are made of other webs, so we must observe them as in Antonioni's film *Blow Up*, finding similarities in differences and differences in similarities. A multidirectional web is an open and centrifugal field in contrast to the closed centripetal sphere.

Theoretical physicist Lee Smolin proposes that the universe itself could choose forms of development of structure and life through evolution as a web of self-organizing systems. (Smolin 1996, 295) He restores the relational view of Leibniz's Monadology over Newton's static perspective on absolute space, stating that: "Quantum gravity, as far as we can tell, goes even further in this direction, as our description of the geometry of space-time as woven together from loops and knots is a beautiful mathematical expression of the idea that the properties of any one part of the world are determined by its relationships and entanglement with the rest of the world." (Smolin 1996, 290)

The most basic self-organizing forms to the most complex take a reticular configuration. At *primus* level, Bénard's hive effect and the tendency of chemicals to be organized in beautiful configurations, as in Belousov-Zhabotinsky reaction, are notable. At *secundus* level increasingly complex orders are organized from nucleotides in DNA replication and amino acid coupling to the narrow organelles and nucleus in symbiotic communities within the eukaryotic cell as stated in Margolis's pioneering work. The animal creature is a web of immune, neurological, endocrine, digestive, circulatory systems interconnected by chemical messengers such as hormones, enzymes, and peptides by chemo-semiosis flowing throughout the body to regulate perception, emotion, memory, metabolism, balance, and health in general. At the next, social scale, the human organism is grouped into webs of communities, languages, social matrices, customs, and civilizations. All these processes are not expressions of hierarchical systems with central control in the brain sending electrochemical signals to control the operation of the individual or social body, but instead heterarchical cross-linked systems running horizontally. Even selection units in gene replication are epigenetic development webs all of which the atomistic vision of isolated genes had ignored until recent research. (Jablonka and Lamb 2005, 280)

This takes us to the view held by biologist Francisco Varela who understood identity as an emergent property of local rules in web patterns, so in his outlook the living world is created in and by the very act of living. Varela reiterated his criticism of the representational and computational view of mind (linear input-output) and proposed instead the concept of *enaction* referring to the subject that creates knowledge by the process of autopoiesis in co-evolution. (Varela 1996, 212; Varela et al. 1992, 151–157, 180–184) In a circular or reticular process, biochemical reactions create a barrier, a membrane in the web, which guards entry to the interior. He concludes that autoimmune diseases like AIDS at the scale of the body, or like drug addiction at a social scale, are network issues that should be addressed as reconnection problems (of the immune system, or of the individual with his family, work, relationships) rather than as they are generally addressed, by extraction, isolation, and confinement.

This delocalized reticular perspective of emerging dynamic patterns finally dethrones the hard Parmenidean spherolatry. When Pascal writes that nature is "an infinite sphere whose center is everywhere and whose circumference is nowhere" he was probably referring to a web.

Signo Ergo Sum

From the intolerable attack on human vanity perpetrated by the Copernican revolution, the scale of our location in the cosmos has been continuously shrinking. Descartes came to our aid with his metaphysical armor brandishing the *cogito* to reinstall our hierarchy elsewhere. This time in a unique dimension, *res cogitans* or thinking thing, with the high-flying company of, no less, God; he thus proclaims that he proves our existence by the simple principle of *cogito ergo sum*. For Descartes, bodies are part of *res extensa* and, as machines (Dawkins winking), obey the same laws of physics that govern all material entities. But he paid the high price for this in the split mind and body, matter and spirit that we have not yet managed to reimburse, and so Descartes was transmuted from the hero of modern rational thinking to the villain of contemporary embodiment thinking. (Damasio 1994)

The Cartesian concept of "extension" was fatally challenged by atomic physics proving that matter is mostly organized as empty space between atoms and particles and, in the case of black holes, almost infinite density without extension. However, this concept of extension seems suddenly to acquire partly new validity, even if not in matter but in Einstein's extended space-time that bends by the mass of matter. Also, while matter for super-string and branes theory would not be constituted by mass, it is however extended as rubber bands or loops to infinity. By this perspective, we could designate space-time as "web *extensa*" (rather than *res extensa*) of extension and expansion processes populated by waves and particles, energy and matter fields in the dynamic equilibrium of the *Orbis primus*.

Cuddled up in front of his stove and locked within the seventeenth century, the chilled French philosopher failed to glimpse from his dichotomy the vast signification activity disseminated between this *res extensa* and *res cogitans*. Such vertiginous activity we can call "web *significans*" and is coextensive with all life processes in *Orbis secundus*, intrinsically linked and dependent on the web *extensa* of the *primus*. I have thus coined this title of *signo ergo sum* or "I mean therefore I am" for denoting the vast *significans* network in an activity both existential and practical. For Varela, the living world is created in the very act of living and for Peirce and Morris we may claim this conception of semiosis as co-extensive with life itself. Consequently, *signo ergo sum* is living by the act of signifying, at any scale, molecular, cellular to individual and communal.

In *Orbis secundus* all live beings require the activity of meaning (emitting or perceiving what is significant) to stay alive. The web *significans* takes Sebeok's "ecumenical Semiotics" literally meaning that "semiotics encompasses the whole of the *oikoumeneé*, the entirety of our planetary biosphere." (Sebeok and Agrest 1975, 181–182) From there it branches out through what has been called "the perfusion of signs" that I define as *significans* network where an act of interpretation is an act of "a-sign-ment," as Sebeok termed it, to raise the interpreted phenomenon to the status of signhood. (Sebeok et al. 1992, 340) Living is equivalent to meaning.

> It is important to note that only living things and inanimate extensions experience semiosis, which thus becomes inspiring and necessary, if not sufficient, of discernible attributes of life. By "living things" we mean not only the organisms belonging to one of the five kingdoms, namely that of Monera, Protoctisa, Animalia, Plantae and Fungi, but also components of parts hierarchically developed, starting with a cell, minimal semiotic unit corresponding approximately to fifty genes, or billions (10^{12}) of atoms intricately organized. (Sebeok 1996, 22)

Therefore semiotics mediates within Cartesian dichotomy between *primus* molecules and their binding in peptides whose molecular configuration is significant in DNA replication processes to form proteins and cells towards the *Secundus*. Through this semioweb, messages of affinity or rejection, nutrients or toxics flow between different organisms. Meaning is involved already in the molecular structure for a ribosome, a ligand for a cell, or an antigen for an antibody through a *significans* web among cells, plants, animals.

Cogitans, in its etymological sense, means "agitated with someone," co-agitated. The *cogitans* web emerges in any signifying relation between an emitter and a destinatary (interpreter). Every creature with a neuron participates in the web *cogitans*, human or octopus. *Cogitans* develops more complex interpretation processes than *significans* because it requires not only membrane terminals of the senses connected to neurons but two individuals in the effort to share a message. Substances such as serotonin, norepinephrine, dopamine, oxytocin, endorphins, and pheromones flow through a neuronal system in a web that may consist of a few neuron cells and nerve terminals to 10 billions of them. They also require communication, namely a threshold that considers not only traces and indicators but intentions of the issuer for the reception of the message.

The existence of web *cogitans* in nonhuman creatures is exemplified by the cunning chromatic and morphological mimic octopus in deceiving predators or the honey bee dance flying in an eight figure to indicate pollen location by the angle at which the two circles intersect.[17] This dance was perhaps the vision that converted Thomas Sebeok into the father of the biosemiotic "ecclesia" (as the vision on the road to Damascus converted Saint Paul into a Christian), from the Greek ἐκ·κλησία meaning "calling out" namely all semiotics fields outside linguistics. Nonetheless, before that the Stoics were already well aware that "animals . . . communicate with each other by means of signs." (Sebeok 1977: 182)

> By the thirteenth century, Thomas Aquinas had concluded that animals make use of signs, both natural and those founded on second nature, or custom. Virtually every major thinker about semiotic issues since, from Peirce to Morris to Thom, and, above all, Jakob von Uexküll, have reaffirmed and generalized this fact to encompass the totality of life. Only a stubborn but declining minority still believes that the province of semiotics is coextensive with the semantic universe known as human culture; but this is not, of course, to deny Eco's dictum (1976, 22) that "the whole of culture should be studied as a communicative phenomenon based on signification systems." (Sebeok 1999, 93)

The web *cogitans* is woven by the flow of signification vehicles that operate in some cases like pheromones (of *pherein* transport and *hormone* stimulating) emitting different types of messages such as aggregation, epideictic or territory markers, alarm, attraction or stimulation, path routing, sexual stimulation, or tranquilizers. A well-known case is bombycol, a pheromone emitted by the female silkworm that touching the male olfactory receptor antenna provokes an immediate sexual response. The African vervet monkey (*Cercopithecus aethiops*) has four or five known communicative sounds: a snake impels a call "*chutter*," if there is a large predatory bird he alerts with a "*rraup*," whereas a small mammal is indicated by "*uh*" or "*nyow*!"(Wilson 1980, 105) Descartes would have classified this monkey as *res extensa*, a mere reflex machine, but his monkey partner understands perfectly that it is an altruistic warning.

There are also individual names in different species or signs of identity which Sebeok (1990, 85) calls "singular proper name" or SPN, for they label each particular individual in its social environment through different channels: smell, visual appearance, sound, including electrical signals. So far we have seen how the webs *extensa*, *significans*, and *cogitans* are woven by complex interactions and meanings that enable the worlds of life, mind, and culture to emerge.

Computo Ergo Sum

Norbert Wiener wrote that "information is information, not matter or energy. Materialism which does not take this into account can not survive today."[18] In his tautology and emphasis, Wiener is not helpful in solving the riddle of what exactly information is. For Roger Schank "information is surprises" and Gregory Bateson defines it as "a difference that makes a difference." (Schank 1996; Bateson and Donaldson 1991, 309) Lloyd and Ng show that information has to do with the condition of matter-energy, for example, the spin of an electron, and they state that "every electron, photon and other elementary particle stores bits of data, and every time two such particles interact, those bits are transformed. Physical existence and information content are inextricably linked." (Lloyd and Ng 2004, 31) A physical system can be explained, according to the authors, using a finite number of bits that each particle in the system acts as the gate of a computer. When particles interact it can cause a change of spin or axis. Lloyd and Ng claim that the universe can thus store 10,123 bits of information.

We must not forget, however, the unwelcome guest of contemporary physics, the observer, to whom bits can be informative. Observers as interpretants are firmly planted on the three orbs: in the *orbis primus* by their physical presence and their observation instruments, in *orbis secundus* by their species-specific bodily conditions that allow their senses to detect only certain kind of phenomena but not another, and in the *orbis tertius* that endows technical detection prothesis and professional codes of interpretation for their observations.

A great concern to physicists is the fact that all information disappears in black holes. Their concern is not so much about the leak of information itself but about the violation of the laws of quantum physics that assumes that information should be conserved according to the general principle of the universe which records and processes information. True to this principle, Stephen Hawking (1977, 2005) claimed that even in black holes some of that information is expressed by radiation, as in the hot carbon. If matter cannot escape the black hole, information could do so, although several theories to explain this case do not seem to have achieved consensus.

A Cartesian dualistic universe would have to be understood either as a monistic *computo ergo sum* (since where there is matter, there is information), or as a trichotomy: proto-semiosis of information at web *extensa*, semiosis of meaning at web *significans*, and semiosis of communication corresponding to web *cogitans*. Information exists in latent form previous to the interpretant as a result of the order of the universe, because in chaos and total entropy, no information is possible. As Paul Davies points out in his Templeton Prize Address:

> To me, the true miracle of nature is to be found in the ingenious and unswerving lawfulness of the cosmos, a lawfulness that permits complex order to emerge from chaos, life to emerge from inanimate matter, and consciousness to emerge from life, without the need for the occasional supernatural prod; a lawfulness that produces beings who not only ask great questions of existence, but who, through science and other methods of enquiry, are even beginning to find answers.[19]

Cogito Ergo Talis Mundus Est

When Hegel formulated the great questions of existence, he explained the future of the world as a result of negations and contradictions in a dialectical process of objectification of spirit or *Geist* by knowing or revealing itself through matter. However, from the perspective outlined here, what we witness is actually the opposite: it is not the spirit but the matter that is diversified to engender the almost miraculous spirit in mental activity whose probability of emergence was near zero. It is a process of objectification, yes, but not of a subject or *Geist* observed in its development from the beginning of the creation by God or a designer of the universe, but a retrospective observer resulting from such process that takes the universe for its object. That is why astronomers today can perform the amazing deed of detecting the Big Bang, whereas the Big Bang could not possibly detect these astronomers.

How is it possible for us to know the universe? Charles Pantin began to think in 1965 on what is now branded as the "anthropic principle" defined by the Oxford dictionary as "the cosmological principle that theories of the universe are constrained by the necessity to allow human existence." Brandon Carter (1974, 294) distinguishes the weak anthropic principle (WAP) implying that, on one hand: "We must be prepared to take into account the fact that our location in the universe is necessarily privileged to the extent of being compatible with our existence as observers." On the other, he proposes the strong anthropic principle (SAP): "The Universe (and hence the fundamental parameters on which it depends) must be such as to admit the creations of observers within it at some stage." To paraphrase Descartes, *cogito ergo mundus talis est*."[20] The phrase "I think therefore the world is as it is" does not mean that I generate the universe by thought but that being the universe as it is, thinking about that universe was possible in a place and time. The universe does not create its observers, but observers are a logical consequence of the structure of the universe at a time and place. We do not know the "why" or "what for" of the universe, only "that" and "what." As two and a half centuries ago Kant wondered about the conditions of possibility of knowledge, such conditions today are given in and by the conformation of both the world and our own in it, even if only in a specific place and time . . . and fleeting.

Hence we can derive an answer to the big question of how it was possible that matter capable of feeling matter would emerge and how sensibility could be attuned to it. Why is it that this matter differentiates itself from the environment in a closing contour while leaving a crack open? The concept of autopoiesis and protobiotic vesicles which are closed on themselves while allowing penetration of molecules and information storage may also be storing an answer.

In this table, we show a relationship between the three world's structures and partially overlapping configurations, whose boundaries are fuzzy.

We can notice that the proportions in this scheme, which only proposes relationships, are totally out of scale, as extension of *primus* in the universe is incomparably greater than *secundus* that, to our knowledge, only exists on

Orbs	*Primus*	*Secundus*	*Tertius*
Configuration	web *extensa*	web *significans*	web *cogitans*
Semiosis	information	signification	communication
Dynamics	inertial motion	effector /perceptor cycle	actor/reactor cycle

Figure 2.1.

planet Earth. Also, *secundus* is incomparably greater than *tertius* produced by cultural species. It should be emphasized, however, that the cycle effector/perceptor as well as signification and cognition occur both in *secundus* and *tertius*. Because orbs are superposed and emerging from the previous background, elements of the preceding network can be found in the following level; but not the opposite. All communication involves signification, though not all signification involves communication. All signification and communication involve information, although not all information is signified or communicated.

Despite attacks wielded against the Cartesian dichotomy, we can eventually rescue the concept of *cogitans* as a category of analysis to refer to the communicative activity and not as an ontological distinct reality. Creatures of different species possess minds in varying degrees of complexity, but communication is a phenomenon of convention present only in social species such as ants, bees, primates, birds, and typically in the homo sapiens, but, until now, we only know that it is foreshadowed in plants when they alert about plagues or create illusory effects of being plagued to prevent leaves from being eaten by insects.

Culture starts from communication and co-agitation. It would be enough to have two individuals in repeated communication and become a habit for culture to emerge, from the songs of a couple of gibbons to the daily life of Robinson Crusoe with Friday. Collective behavior exists already in the intonation of bacteria like *Vibrio fischeri* when as soon as they perceive the proximity of a critical mass through quorum sensing, they ignite together.

While the scientific method fails to explain how subjective experience is possible or why it feels as it feels, aesthetic and cognitive theory remain feasible routes to understand these phenomena. It is not only a question of explaining consciousness and distinguishing between "mind" and "brain," but also digging into the enigma of qualia. It's still a mystery how electrochemical activity between neurons is related to dreams, mental images, and reality models that represent them. Bateson mentioned, I can't recollect where, that thanks to this ignorance on how we produce mental images, we can trust them and their supposed correspondence with reality for survival purposes.

The conditions, reasons, or explanations for these huge leaps in level between the three orbs have been part of cosmological myths, religions, and sciences. Jamsa writes that evolution is like an unscripted theater play always improvised in retrospect that seems to tend to complexity and, we might add, towards aggregation processes also.[21] Mutations or irreducible qualitative leaps occur where each order emerges from a greater quantitative world generating greater qualitative complexity and demarcation. As the evolution of *primus-secundus-tertius* moves towards a progressive order and differentiation it definitely appears to be a process opposite to entropy.

ENTROPY AND HERESY

Are utzijoxik wa'e, k'a katz'ininoq, k'a kachamamoq, katz'inonik, k'a kasilanik, k'a kalolinik, upa puch katolona kaj[22]

—*Popol Vuh*

One of the fundamental principles of the *orbis primus* is entropy or the second law of thermodynamics, which states that in a closed system matter and energy tend to homogenization or disorder since the quality of energy is irreversibly degraded. This sacred law of the *primus* seems violated by the *secundus* because matter and energy, rather than degrading, tend to heterogeneity, order, and progressive differentiation through the emergence of a variety of species. This violation of entropy Schrödinger (1962) called "negentropy" (negative entropy). The negative entropy directly involves the proliferation of semiosis, because every sign is unlike any other sign and keeps generating further order. Sebeok states that:

> Schrödinger's discussion indicates the relevance of semiotics in understanding the processes of life, or, as expressed by Wiener (1950, 21)—given that the common view that the object of semiotics is the exchange of messages (i.e., time series)—the amount of information is a measure of "the degree of order that is particularly associated with the patterns that are distributed in the form of messages in time."[23]

For Sebeok (1999), another significant feature arises from the interaction between the essential invariance of biochemistry underlying life and the lavish variability of singular embodiments thereof, in parallel with the global semiotic universals and so-called local or cultural variations.

Everything grows by self-organization through semiosic action from the simple to the complex in a progressive order of diversifications and bifurcations. "Symbols grow. They come into being by development out of other signs. . . . It is only out of symbols that a new symbol can grow. A symbol, once existing propagates through symbols. For every symbol is a living thing, in a very strict sense that is no mere figure of speech."[24] From semiosic growth, everything else grows, as in Raffi's nursery rhyme: *Everything grows. Babies, animals also. Everything grows and grows, sisters, the brothers also, everything grows, the food at the farm, fish in the sea, birds in the air. . . . A blade of grass, toes on my feet, finger in my hand, hair in my head, a red, red rose.*

A consequence of entropy, if one principle is applicable to the three orbs, is irreversibility, but it operates differently in each of them. In the *primus*, entropy prevents a broken cup to recompose or poured water to spontaneously return into the container. In *secundus,* there is no return from death. The central dogma of molecular biology states that information can move from DNA to proteins but not vice versa, hence the impossibility of hereditary acquired characteristics as, say, inheriting the understanding of the theory of relativity or the ability to play Chopin's #6 prelude. However, mutations for antibiotic resistance among bacteria are both acquired and hereditary, a triumph for Lamarck. Swenson (1999) and Kaila and Annila (2008) propose that this apparent violation of entropy in *secundus* can be explained due to the fact that living systems are the most efficient paths for energy dissipation and thus the second law eventually remains in *orbis secundus*. Regarding *tertius*, things get even more complicated, because where is the entropy in culture when we testify in scientific paradigms the impressive diversification of knowledge and progressive construction of scientific order?

We cannot deny, despite the pessimism that may overwhelm us, that there is a degree, even if small, of evolution in culture when freedoms are won, mysteries are clarified, injustice fought, sometimes.

In our terms, the semio-web interlaces at all levels for the last 4 billion years on earth through continuous processes of action-perception between sensor and effector in von Uexküll's functional cycles. Evolution depends on semiosis for detection and absorption of solar energy, for nutrients and water by plants, of vegetal food for animals that in turn will feed fungi, bacteria, and animals. The most heretical creatures that contravene entropy are plants when they perform the miraculous feat of breaking the compact molecules of water and carbon dioxide to form glucose and release oxygen into the atmosphere. They convert solar energy into chemical energy from which all the rest of creatures feed.

In the previous section I mentioned that *signification* corresponds to *action* as *information* corresponds to *motion* of quantum or celestial bodies in the *primus*. There is no *action* of the sea but the *motion* of the sea waves by the proximity of the moon to the earth; there is no *motion* of pseudopodia, but an *action* of the amoeba to envelop nutrients. Physics deals with *motion* and its laws, causes and effects, not with *actions*, intentions, reactions, responses or interactions that are characteristic of *secundus* and *tertius*. Newton's third law, "to every action corresponds an equal and opposite reaction," is better phrased: to every motion corresponds an equal and opposite countermotion.

Unlike motion in the *primus*, action in *secundus* is always intentional (oriented) at the three levels of autopoietic systems from cell to multicellular organisms to social groups. In action there is intentionality while in motion there is causality. Between these two extremes, everything is transformed by a continuous process of subjectification and objectification, perception and action in functional cycles. Motion can be explained by physical, thermodynamic, and chemical laws; while action can be explained only by biosemiotic laws because there is no action without meaning.

From Thure von Uexküll's distinction between exo and endo semiosis, we could say that every autopoietic unity at any level performs two classes of semiosis: 1) endosemiosis in metabolism, growth, and reproduction within the cell, 2) exosemiosis by coordination and diversification in relation to a particular node through the semiosic web among various cells or between various components of a same cell (i.e., mitochondria or Golgi apparatus in relation to the cell nucleus). (T. von Uexküll, Werner, and Herrmann, 1993)

The eukaryotic cell, discovered by Margulis as a result of coordination in symbiogenesis, is one example of endosemiosis, as in DNA replication during cell division, while exosemiosis regulates epigenetic differentiation of daughter cells according to their position within the embryo, implying a horizontal transmission. The code used during DNA replication is digital in endosemiosis while, during epigenesis, probably an analog code is at work by mimicking the neighboring cell morphology in exosemiosis.[25]

Actions and reactions are found only in live beings who are actors and reactors, at the second level of the functional cycle in semiosis. The effector-perceptor relationship proposed by von Uexküll is to endosemiosis as the actor–re-actor I am proposing here is to exosemiosis, since it addresses another subject. Any living system performs specific actions related to its survival and always acts in relation to meaning. Put in other words, every relation is a meaning relation. As observers, we can determine the variety of processes of signification and meaning precisely through the variety of activities undertaken.

EVOLITION

One of the anomalies in the theory of evolution by random variation and selective retention is, as we just asserted, that it generates progressive diversification, order, and complexity, challenging the second law of thermodynamics through what Schrödinger defined by the term of "negentropy." Neither Darwin nor neo-Darwinism intended to explain directionality in evolution and simply assumed natural selection of mutations as a fact, exploring its consequences and heuristic potential.

On the other hand, if it were true that evolution has no direction or meaning, what explains such variety, change, and irreversibility? What impels us to improve when it is more economical to repeat by habit, why not abide by the least effort and remain the same in perpetual homeostasis? Why are species not fixed, as in Noah's ark? In sum, why does nature evolve? Why does culture? And a nonscientific question: why evolution at all?

Despite Gödel's theorems on the impossibility of a theory being both consistent and complete, scientists insist on pursuing such possibility, at the expense even of recruiting the idea of God as the great solver of all riddles, who is actually the great dissolver of all scientific consistency. Newton made this move regarding the origin of planetary motion, which nevertheless did not compromise the accuracy of classical mechanics' three laws of motion and universal gravitation. However, not everyone was that lucky.

Among those who appealed to theological principles literally or metaphorically for explaining problematic aspects of evolution we must mention Alfred R. Wallace, Pierre Teilhard de Chardin, Ludwig Klages, Gregory Bateson, Stuart Kauffman, Jakob von Uexküll, and Humberto Maturana.

Bateson spoke of the "sacred," "spirit," or "ecological mind" of living systems (similar to what von Uexküll called "master plan") which refers to evolution itself as a mental process from a horizontal, immanent more telluric perspective. Such mind happens to dominate nature by means of culture through agriculture, technology, medicine, and education in the same way as nature dominates the inanimate world when plants absorb energy from the sun and break molecules in water for nourishment. Furthermore, organic replicating DNA dominates the inert molecular world by imposing its own structure, complexity, reproduction, and dynamism to amino acids and other elements. This directionality is leaving an irreversible trace and, perhaps also, creating time.

In this section, we will consider another case, that of Friedrich S. Rothschild, who not only appealed to but also developed a theological perspective of evolution as inner adaptation within transcendental subjectivity to an omnipotent and omniscient God. This particular perspective would not have theoretical relevance for biosemiotics except for the fact of focusing upon two aspects that evolutionary theory has paid little attention to, namely the question of directionality and of subjectivity. This double aspect, I will contend, raises deep and important questions that need to be addressed, especially if we pretend to center our attention on aesthetics as a theory of sensibility.

Biosemiotics and Subjectivity

As in the almost coincident invention of photography by Niépce in 1826 and by Hercule Florence in 1832 or of semiotics at the turn of the twentieth century by Peirce and semiology by Saussure, another concurrence took place in the study of signification among various species simultaneously proposed as "biosemiotics" by Friedrich Rothschild in 1962 and as "zoosemiotics" by Thomas Sebeok in 1963.[26] The line of work of these two authors is, however, considerably divergent. Sebeok, a semiotician self-portrayed as a "frustrated biologist," defines semiosis as equivalent to life and proposes the study of sign systems in all animal species.[27] Semiotics for Sebeok is a matter of communication which necessarily implies subjects exchanging information. The subjective dimension inevitably emerges in correlation to semiosis.

Rothschild, a neurologist, psychiatrist, theologian, and, we could say, frustrated semiotician, considered the various stages of evolution from the perspective of meaning and understood communication as the meaning-carrying axis of the evolutionary process. He defines biosemiotics as the "psychophysical link in the central nervous system and in structures that possess psychophysical functions in organisms."[28]

For Rothschild, also, semiotics always involves communication and therefore implies subjectivity in sign-exchange processes. "This subjectivity is certain, although we cannot observe it objectively. It is obvious to us in the expression. It is a mediated directness of the co-experience with the subjectivity of the other, a connection between one subjectivity and another. Mediated directness of the connections is the principle dominating all our relationships to our subjectivity." (Rothschild 1994, 20) He proposed to call this branch of science bio-semiotics.

Up to here, both Sebeok and Rothschild appear to agree in the saliency and necessary condition of subjectivity for semiosis—an aspect present in Peirce's interpretant but absent in many others (i.e., Saussure's positivistic semiology).

In the effort to understand the evolutionary integration of biological and cultural worlds, Rothschild notes that "protozoa, invertebrates, vertebrates and finally man appear as four stages of development of subjectivity. In each one, a new sign system overcomes the existing ones and performs the unfolding of a new and broader level of possible experience."[29]

He proposes the evolution process from a nuclear system to cellular, gastrular, neurular, and noetic. The gastrular system imposes directionality to the body, from mouth to anus, and consequently a corporeal interiority (as the neurular enables a perceptual interiority through the senses). He notes that each new system must be related to the previous ones through a qualitative affinity and must also adapt to the previous before generating more complex structures. (Rothschild 1994, 67–69)

He summarizes in three of what he defines as the laws of biosemiotics which, thanks to Kalevi Kull, I came to read: 1) intentional action for the conservation of structure, 2) polarization and asymmetry, and 3) rule or law of transcendence and historicity of semiosis. I quote verbatim:

> The first law.
>
> Threat to given life elicits from the original passive state of the organism a component of activity, of inner self-assertion, transforming it from an object into a subject of intentionality. The first biosemiotic law expresses the intention to safeguard the structure that conveys the own essence, the self as a coherent unity. It is the basic rule of biosemiotic syntax.
>
> The second law.
>
> Inner polarization is necessary in order to permit the subjectivity of organisms to communicate with the objects of the world simultaneously with realization of the own self. This law dominates the arrangement of all communication systems from the cell upward. The manifestation of this inner polarity include the differentiation of motor and sensory systems in the sensori-motor foundations of experience and behavior, the bisexual disposition of organisms, the asymmetry between right and left, the differentiation of the vegetative nervous system into a parasympathetic and sympathetic component, and the arrangement of the central nervous system in homolateral and heterolateral centers.
>
> The third law.
>
> As each new inner communication system emerges in evolution, it transcends its predecessor's horizon of meaning and requires for its actualization a new mode of intentionality. In this new form of intentionality, subjectivity is active and dominates over that of the preceding system because it is in opposition to it and thereby prevents an independent activity of the more archaic systems. The necessity of this dominance constitutes the third biosemiotic law: without such dominance, the new system cannot develop its function. (Kull 1999, 779–81)

In the first law we recognize evolution and autopoiesis as guidelines. Subjectivity emerges in the drive to survive, because what evolves is not only the morphology of creatures as individuals separate from the environment, but their modes and tools of perception, including self-perception. This quest for survival implies directionality and consequently choice. The second law relates to the separation of input and output in the metabolism or replication from parthenogenesis to sexualization. Such asymmetry allows integration to more complex levels enabling the distinction between perceptor-effector processes, male-female, self-nonself, inhaling-exhaling, ingesting-excreting and between the two cerebral hemispheres or cell's nucleus and cytoplasm. The third law allows us to take into account evolution from the inert to the animate to the mental in terms of a hierarchical development in their structure and subjectivity.

As stated by Sebeok quoting Shrödinger and Jakobson: (Sebeok 1999, 87)

> There are several additional noteworthy properties of life. One of these is its hierarchical organization, "a universal characteristic which life shares with the rest of the cosmos and which defines, in the overall architecture of the universe, its position on the genealogical tree." The hierarchy of nature appears as an ontological interpretation of data from the "real world," a pattern of relations which obviously extends up through semiotic systems, including particularly the verbal. (cf. Jakobson 1960)

On Meaning and Value

The various meanings of "meaning" cover the axiological, semantic, epistemological, normative, pragmatic, psychological, existential, theological, and teleological. Rothschild quotes Victor Frankl for whom meaning is justification or mission in life and becomes almost synonymous to value. According to Frankl freedom consists of endowing life with meaning, so that the inability of the will to find meaning is a factor that leads to neurosis. Nevertheless, is one free to give life a meaning and not another, or not give any meaning at all? Sam Harris claims that it is not we who are deciding to decide nor who decide to act on what we decide. There are neural processes operating at a level below consciousness which decide courses of action before they appear to consciousness. "My choices matter—and there are paths toward making wiser ones—but I cannot choose what I choose. And if it ever appears that I do—for instance, after going back and forth between two options—I do not choose to choose what I choose. There is a regress here that always ends in darkness." (Harris 2012, 8)

Rothschild challenges Darwinian theories for their disregard of evolution's sense and direction and proposes to establish a science that deals with the meaning of evolution. For Rothschild, meaning is a matter of inner adaptation to God: this implies that individual subjectivity is open and receptive to Transcendental Subjectivity in its dynamic materialization of it. One may assume that as a religious man he had the answer before he even made the question and that Darwin never pretended to give any meaning to evolution but simply explain its mechanism. Yet, there is a point worth taking before dismissing this problem altogether as a matter of belief. Again, it is the question of directionality we have no reason to take for granted.

Quoting Rothschild (1994, 55): "Being comes into existence as expression of meaning. Meaning is immaterial and full of intentions that will become manifest in the freedom of thought and volition of mankind." Rather, the opposite seems exactly to be the case, namely, it is meaning that comes into existence as an expression of being, since it is the living creature (and not only humankind) that values and gives meaning to the events in its surroundings and which is "full of intentions." In the drive for survival, whatever affects it has meaning and value to an organism. Consequently, all creatures evolved and perfected adequate sensors to signify such meanings and to assess such values. Here again we can fully attest the intimate link between aesthesis (in perception) and semiosis (in signification).

From a semio-dynamic approach, Alexei Sharov raises the distinction between meaning and value in these terms: "Meaning is a set of bans and limitations set by information on the trajectories of system development and behavior, and value is measured by the contribution of information to the safety of self-maintenance and self-reproduction of the system." (Sharov 1992) In Sharov's distinction, those bans and limitations of meaning seem to be a result of differences and oppositions among positions within that same system. Meaning marks the direction that such a system takes at each crossroad of possibilities. Consequently, meaning is the activity of the system at each bifurcation.

Value, on the contrary, is intrinsic and pragmatic, while both are totally affected by the context. Sharov understands that: "subjective evaluation is only one specific sort of values—the ideal value. But besides there is material value expressed in the ability of systems to maintain and reproduce themselves." (Sharov 1992) This dualistic distinction between material and ideal, however, is untenable since value is always material as it affects a creature's survival.

Directionality in Evolition

Evolution is not going anywhere in particular, but comes from a particular set of conditions; hence by hindsight we can become aware of a clear directionality. The power to choose, to be fruitful and multiply reveals that evolution is not in the hands of omnipotent, omnipresent, and omniscient gods, but instead depends on minipotent, minipresent, and miniscient creatures oriented to survival and reproduction. It is we, macro and microscopic creatures who create meaning and are created by it. From slime mold forming pistils in times of hardship to disperse spores that provide a better opportunity for their species, to yeast that selects only one companion for life, the future of our species is set by selecting every mark and every trace in every bifurcating path of evolution.

At this point, Rothschild's orthodox theology paradoxically coincides with Harris's orthodox atheism, coming from opposite directions: the former by prospective or foresight and the latter by retrospective or hindsight. For Rothschild, the Transcendental subject has the route already marked to which the subject must adapt in the process of evolution and improvement of the world. His will is adapting to the divine. For Harris, developments and circumstances have marked beforehand all what supposedly he could choose. "Some moments before you are aware of what you will do next—a time in which you subjectively appear to have complete freedom to behave however you please—your brain has already determined what you will do. You then become conscious of this 'decision' and believe that you are in the process of making it." He adds that "the intention of doing one thing and not another does not originate in consciousness—rather, it *appears* in consciousness, as does any thought or impulse that might oppose it." (Harris 2012, 8–9) This echoes James's (1884) idea that it is the body that reacts directly to perception before the meaning of the situation is processed by the mind's awareness.

We must here distinguish between consciousness and will. That impulse toward action arises from the body, as we have seen from James's pragmatist psychology, does not imply a lack of the activity of decision or canceling the operation of the will, nor is it required that it be conscious. The choice between one partner and another is a bifurcation point that alters the trajectory of a species and the significance of phylogeny. The more complex the organism, the more choice and freedom it has. Survival defines the direction in selection, values, and meanings of our actions from the perception of each horizon. The evolution of will and the will itself in evolution through every crossroad involving freedom and subjectivity is precisely what we mean by the concept of *evolition*.

Evolition entails that each creature in its own situation can choose the next step in every bifurcation according to the value it conveys to each alternative. The sum of these choices taken by all creatures based on meaning and value establishes directionality to evolution from a retrospective view.

This has resulted in the conflation between a teleological notion of adaptation (e.g., creatures evolve in order to adapt to their environment), a theological notion of adaptation as in Rothschild (creatures are created to adapt to their creator), and a functional idea (adaptations are phenotypic results of natural selection) where only the latter is scientifically correct. Briefly:

1. Evolution is not a process for adapting to environments. We do not evolve in order to adapt, but adaptations are results of interactions with environments. Adaptations are not teleological and prospective, nor theological (as Rothschild's adaptation to God) but instead the result of a retrospective view by applying reverse engineering to understand phenotypical features.
2. We cannot speak of meaning unless there is freedom. In a deterministic world, there is no meaning since there is no choice (i.e., Sharov's example of the life convict for whom information about the outside world has no meaning).
3. Choice and value are co-dependent, and evolutionary directionality follows value at bifurcations.

4. Evolition is the subjective dynamic resulting from meaning and value that establishes evolutionary directionality.

In the evolutionary theology of Abraham Isaac Kook (who probably influenced Rothschild's work): "Each creature has, in accordance with its ability, a part in the choice, and that is the basis for its betterment in the future." But the category of "choice" (adds Yosef Ben Shlomo), applies in reality to *everything*, and it is expressed in the gradual particularization of the world. "Every stage in its development and progress 'chooses' its special place on the ladder of the stages of existence, in a good and just choice of the cosmic will which acts within it." (Ben Shlomo 1990, 61, 64)

Rothschild ventured through a path so entangled in theology that it comprised risks similar to those encountered by the four sages who entered the *pardes* (orchard)—Ben Azzai, Ben Zoma, Acher, and Akiva. According to the Arameic Talmudic legend Ben Azzai gazed and died; Ben Zoma gazed and lost his mind; Acher destroyed the plants; Akiva entered peacefully and left peacefully. Perhaps the path that Akiva found out of this orchard was simply following the trail of the smallest of all creatures.

VIBRATING

Rhythm makes the universe intelligible. Pythagoreans imagined the music of spheres as a bridge between the stars and men. Rhythms in the *orbis primus* mark spatial and temporal cycles in planets' translation and rotation orbits. Newton saw rhythms also in waves emitted by the light spectrum through the glass prism. Color is undulating rhythm and with it we can measure not only distance but time according to frequency range in light waves by astronomical spectroscopy. Dawkins devotes a book to this wonderful symphony of rhythms in the cosmos that can be decoded through the Fraunhofer spectrum to record the presence and distance of stars and other objects in the universe. (Dawkins 2000, 50) The folds of rhodopsin molecules and DNA helical curve display a clear rhythm. The chemical regularities in Belousov-Zhabotinsky reaction, the recurrences in quasicrystals and the entire fractals, the cadences of plant and molluscks' growth all express a spectacular sense of rhythm.

In *orbis secundus* rhythms in seasons, breathing, maturing, flowering, and fertilization or circumnutation regulate plant life. Circadian, digestive, menstrual, inhalation and exhalation, sexual, cardiac, and generational rhythms order animal life. The embryo feels the rhythm in her mother's pace, her heart, and in the inflection, volume, and tone of her voice. By rhythm, we can visualize the brain in the act of thinking (fMRI), the fetal heartbeat by ultrasound, and the structure of our internal organs. Shivering, chewing, walking, running, caressing, and swinging are rhythmic. Making love is rhythm. Dawkins mentions various types of rhythms and cycles such as quadrennial animal population groups and the relationship between predators and their victims. He notes that even biological rhythms of 26 million years of mass extinctions in some species apparently result from celestial cycles. (Dawkins 2000, 74–6) "The rate at which the nerve impulses arrive—which may be hundreds per second—is a coded representation of (in this case) the intensity of light falling on the rod or cone cell. As far as a single nerve cell is concerned, the difference between strong stimulation and weak is the difference between a high-speed machine gun and intermittent rifle fire." (Dawkins 2000, 56)

The singing of whales, mice, or nightingales follows rhythms that the human ear has to translate to perceive more fully. Stickleback fish swim rhythmically in zigzag during courtship, monkeys move their tongues rhythmically when courting, and chimpanzees sway from side to side. "From bears to beetles, courting couples perform rhythmic rituals to express their amorous intentions" notes Helen Fisher (1992, 31). The stridulation of insects and birds singing or drumming to attract females of *Cicada septemdecim* are among the thousands of rhythmic manifestations of the animal kingdom.

> The perception, if not the enjoyment, of musical cadences and of rhythm is probably common to all animals, and no doubt depends on the common physiological nature of their nervous systems. Even Crustaceans, which are not capable of producing any voluntary sound, possess certain auditory hairs, which have been seen to vibrate when the proper musical notes are struck. It is well known that some dogs howl when hearing particular tones. Seals apparently appreciate music, and their fondness for it "was well known to the ancients, and is often taken advantage of by the hunters at the present day." With all those animals, namely insects, amphibians, and birds, the males of which during the season of courtship incessantly produce musical notes or mere rhythmical sounds, we must believe that the females are able to appreciate them, and are thus excited or charmed; otherwise the incessant efforts of the males and the complex structures often possessed exclusively by them would be useless. (Darwin 1871, II, 333)

Brian Goodwin understands life as a particular state of organization, a physical and chemical system whose rules of formation we must decipher. He stresses a chronobiological vision where organisms are rhythmically structured entities that unfold in space obeying the laws of physics and morphology as in a dance.[30]

The *orbis tertius* is regulated by rhythms in festivals, dances, rites of passage, social cycles, school and family periodical meetings, recurring religious ceremonies, political repetitive slogans, and alternation of labor and leisure time. Language, in turn, consists of rhythms, starting with the acoustic and phonetic range of each language. "Theoretically, there is a continuous spectrum of vowel sounds. Any one language employs a useful selection, a discontinuous repertoire picked out from the continuous spectrum of available vowels. Different languages pick out different points along the spectrum." (Dawkins 2000, 78) Hence changing phonetic ranges in mother tongue becomes almost impossible, as certain sounds are unpronounceable after a linguistic range is set. Whoever grew up in Hungarian or French phonetics, bears its vowels and consonants forever.

> [W]hen vivid emotions are felt and expressed by the orator, or even in common speech, musical cadences and rhythm are instinctively used. The negro in Africa when excited often bursts forth in song; "another will reply in song, while the company, as if touched by a musical wave, murmur a chorus in perfect unison." Even monkeys express strong feelings in different tones—anger and impatience by low,—fear and pain by high notes. The sensations and ideas thus excited in us by music, or expressed by the cadences of oratory, appear from their vagueness, yet depth, like mental reversions to the emotions and thoughts of a long-past age. (Darwin 1882, 571–572)

To know is to find the rhythm and order of phenomena, a relationship between movements or actions that maintain regularity in time and space: the contraction and dilation of the heart and lungs, the recurrent beat of masons while laboring, the orbit of a planet, the staccato of a pianist, the Sequoia wood configuration, Van Gogh's brushstrokes, Fibonacci scale in snails' shells, the pendulum clock movement, the cycles of ripening peaches, paddling a canoe, a canary singing, lulling a baby. The allure of prayers is in rhythm. In the embryo all cells divide at the same time and the same rate up to 2000 when they cease to be synchronous. (Hallé 2002, 86) The rhythmic and synchronous fireflies light up the trees of the forest and their light is so powerful that it can guide navigation from the port.

Our universe is undulating and rhythmic in *primus*, everything grows in *secundus*, and everything vibrates in the three worlds: heat, radar, laser and microwave, sounds, voices, superstrings, flowers, and songs. There is rhythm in conversation, in fractals, in our ribs, in the waves of the sea, in the roaring wind and clouds setting in a whirlwind and the chirping of crickets, in water ripples and croaking frogs, in artichokes, in crying, in intercourse and prayer. If God exists, She must be pure rhythm.

NOTES

1. According to my best understanding of Hawking (1996).
2. Jorge Luis Borges, T*he Analytical Language of John Wilkins* can be read online at https://ccrma.stanford.edu/courses/155/assignment/ex1/Borges.pdf. Accessed June 16, 2014.

3. We could find endless taxonomies, such as the Green Plant Phylogeny Research Coordination Group. In 1999, members of the Deep Green Project classified plants in four: plants and green algae, red sea plants, brown sea plants, and fungi. Cf. http://ucjeps.berkeley.edu/bryolab/GPphylo/. Accessed June 16, 2014.

4. Foucault (1986), Preface.

5. Initially I considered the term "pleroma," (taken from Gregory Bateson) which is derived from Gnosticism, and designates the fullness, to describe the inorganic material universe. Jung in his anonymous and posthumously published Septem Sermones ad Mortuos uses this term to contrast it with Creatur but in a totally different sense to that proposed by Bateson. (Bateson 2000, 26–31) In previous works I called the *Orbis secundus* "Biome" and *Orbis tertius* "Culturoma." (Mandoki 2009) I finally opted for the simplest terminology as first, second, and third world in Latin to avoid confusing daily use of the words for the analytical terms to denote the physical, biological, and cultural.http://www.faroaustral.com.ar/wp–content/uploads/2007/06/jung–carl–gustav–siete–sermones–a–los–muertos.pdf.

6. In the distinction of work and labor according to Arendt (1998) passim.

7. Quoted in Abbagnano (1963, 434–35).

8. http://www.sciencedaily.com/releases/2007/04/070419115551.htm. *ScienceDaily* (Apr. 21, 2007) Accessed June 16, 2014.

9. http://www.sciencedaily.com/releases/2010/06/100621101420.htm. *ScienceDaily* (June 22, 2010) 14/07/2010.

10. Mentioned in Hawking (2002, 54, 118).

11. Cf. Mukerjee (2003).

12. I am referring to Sebeok and Umiker-Sebeok (1986).

13. Compare to Sebeok and Umiker-Sebeok (1987). See also Kull (1998).

14. Quoted by Lewontin (2001, 4).

15. My translation from Spanish in Borges's story, which can be read at this site: http://mycours.es/gamedesign2012/files/2012/08/The–Garden–of–Forking–Paths–Jorge–Luis–Borges–1941.pdf. Accessed June 16, 2014.

16. Translation mine.

17. Von Frisch (1954) and (1950). Wilson notes that the value of communication is enhanced by the fact of bits gradation, as the volume, because it gives much more information in discrete units. (Wilson 1980, 96)

18. Quoted in Rothschild (2000, 104).

19. http://www.firstthings.com/article/1995/08/003–physics–and–the–mind–of–god–the–templeton–prize–address–24. Accessed June 23, 2014.

20. Cf. Carter (1998, 133, 135) in Stenger (1999), Uršič (2002), Hawking (1996), Aviezer (1999), and Rees (1996, 266, 271).

21. Jamsa in Barbieri (2008a, 79–80).

22. "This is the account of how all was in suspense, all in calm, silent, motionless, beating and empty expanse of the sky." In Ki'che' a dialect of Mayan language.

23. Quoted in Sebeok (1990, 87).

24. Sebeok (1977, 181), quoting Peirce (2.302), (2.222); Merrell (1996).

25. Cf. Code duality hypothesis (Hoffmeyer and Emmeche 1991).

26. Cf. Rothschild (1994) and Sebeok (1963, 1965).

27. Sebeok (2001, 61), Sebeok (1995), and Sebeok, Umiker-Sebeok, and Young (1992).

28. Quoted in Anderson (2003).

29. Rothschild in *Annals of New York Academy of Sciences* (1962, 775) cited by Kull (1999).

30. Goodwin in Brockman (1996, 103).

Chapter Three

Orbis Secundus

Omne vivum ex vivo

In an infinitely small area of the *primus* a prodigious arrangement of molecules is nested incubating the *orbis secundus*. Unlikely and practically impossible, a particular alignment of carbon atoms favored the materialization of longer, more stable molecular chains, more flexible and with greater combinatorial potential. Thus, a whole new universe emerged: life. Perhaps pointing at this contingency Monod referred to when saying that evolution is chance taken by the wing.[1]

Crystals illustrate the perfect order in the *primus*, but it is a dead order. The beautiful alignment of molecules in the Belousov-Zhabotinsky chemical reaction or the hive effect pointed by Henri Bénard (hexagon alignment of a layer of molecules on the surface of a liquid when heated below) suggest possibilities of how a different order can spontaneously arise in the *secundus*.

From the theory of complex systems, Stuart Kauffman points at two mechanisms to explain the emergence of biological reality: self-organization and evolution, in that sequence, since self-organization precedes evolution and imposes conditions upon it. (Kauffman 1993, 120) Evolution does not start in organisms but in molecules that acquire conformations by self-organization on which natural selection operates in their space as fitness landscapes. (Kauffman 1993, 171) The main thesis for Kauffman is that the order usable by selection is abundant, so the emergence of life was not as improbable as we suppose, since it is a very common property of complex systems. (Kauffman 1993, 235) "Life, I suggest 'crystallizes' in a phase transition leading connected sequences of biochemical transformations by which polymers and simpler building blocks mutually catalyze their collective reproduction." (Kauffman 1993, 287) This autocatalytic polymer can be seen as a thread of symbols or, in semiotic terms, as a syntagm, since a protein is a thread with 20 types of amino acids. Aleatory grammars are constituted where symbol threads act in threads of symbols to generate other symbol threads. (Kauffman 1993, 369) Kauffman considers it more likely that life would descend from these self-replicating peptides than from naked gene templates. (Kauffman 1993, 340–1) Against reductionism of biological phenomena in physics, Kauffman jumps to propose unpredictable creativity in the universe that is linked to the sense of the sacred. There he will converge with Gregory Bateson, Alfred Wallace, Jakob von Uexküll, Friedrich Rothschild, and Rabbi Kook who, from very different directions and angles, reached a limit beyond which they glimpsed only the dimension of the sacred.

For Kauffman evolution starts from these molecules competing for survival by natural selection. This thesis does not meet the oparinists' conviction that life's primordial soup was cooked in a pot with light and without oxygen. The Miller-Urey experiment reproduces these original conditions on earth according to Oparin-Haldane and demonstrates that it is possible to synthesize organic (obtaining 22 amino acids) from inorganic compounds (water, methane, hydrogen, ammonia) in conditions simulating a reducing atmosphere and impact of lightning. What we do know is that such soup is cooked with very select molecules.

THE APPARITION OF AESTHESIS

> Every living creature is in fact a sort of lock, whose wards and springs presuppose special forms of key—which keys however are not born attached to the locks, but are sure to be found in the world near by as life goes on. And the locks are indifferent to any but their own keys. The egg fails to fascinate the hound, the bird does not fear the precipice, the snake waxes not wroth at his kind, the deer cares nothing for the woman or the human babe.
>
> (James 1884)

How to define life? For Erwin Schrödinger it is the property of self-assembly against the tendency toward disorder and entropy.[2] Gerald Joyce of NASA defined it as a self-sustaining chemical system capable of Darwinian evolution.[3] From a cybernetic perspective, Bernard Korzeniewski understood it as a web of feedback mechanisms. (Korzeniewski 2001) For Szoztak, "Defining life is notoriously difficult; its very diversity resists the confines of any compact definition. An operational approach focuses on identifying simple cellular systems that are both autonomously replicating and subject to Darwinian evolution." He adds: "[w]e can consider life as a property that emerges from the union of two fundamentally different kinds of replicating systems: the informational genome and the three dimensional structure in which it resides." (Szostak, Bartel, and Luisi 2001, 387) From the perspective I am taking in this inquiry, it will be clear that life is aesthesis, matter perceiving matter and thus ceasing to be only matter.

The liminal region between *primus* and *secundus* could have arisen, as Szostak suggests, from fatty acid chains that close up on themselves and form vesicles that comprise polymers with information. They perform processes similar to metabolism when they assimilate other vesicles surrounding them that have less volume and then replicate polymers within, and so also replicate their information. The more we decrease the scale, the more blurred the boundaries between *primus* and *secundus* appear.

In both senses, as appearance and as apparition, aesthesis governs the realm of appearances. Two movements make aesthesis possible: stimulus or source detection by receiver and linking it or latching on to a meaning. When this bonding to meaning occurs, matter is no longer only matter and semiosis is set in motion. Perception and signification, aesthesis and semiosis, allow the body to open up and distinguish between self and other by autopoiesis and decide to approach or distance itself, to absorb nutrients by attraction or evade predators and toxic substances by repulsion. The body cyclically performs aesthesis by opening toward its object, and also performs semiosis by making distinctions: it also performs praxis by operating and sustaining an action toward a certain goal. It detects its object by aesthesis, signifies it in semiosis, and decides action in praxis in a functional cycle effector-perceptor. Rothschild's three biosemiotic laws are thus expressed: intentionality toward the stimulus, the polarity between aesthesis-semiosis, attraction or repulsion, and the mastery or power of the self in praxis.

The semiotic definition of the sign as *aliquid stat pro aliquo* or something that stands in place of something, translates for aesthesis as *aliquid stat cum aliquo*, where something stands with something and is related to it, the apparition of something for someone. When a membrane receptor detects something there, semio-aesthesis is initiated in firstness (in Peirce's terms). The organism realizes something or *aliquid* (in firstness) and relates it with something else *aliquo* (in secondness). The stimulus or *presentamen* occurs by aesthesis to the interpretant even before the emergence of the sign or *representamen* (again in Peirce's terms) occurs in the process to semiosis.

Humphrey called "thick moment of consciousness" what I call aesthesis as the beginning of subjectivity.

> Surely, consciousness can exist at a much lower level, exist unreflected on, just as experience of raw being: as primitive sensations of light, cold, smell, taste, touch, pain; is the is-ness, the present tense of sensory experience, which doesn't require any further analysis or introspective awareness to be there for us, but is just a state of existence. . . . I call it the "thick moment" of consciousness. What matters is that I feel myself alive now, living in the present moment. What matters is at this moment I'm aware of sounds arriving at my ears,

> sight at my eyes, sensations at my skin. They're defining what it's like to be me. The sensations they arouse have quality. And it's this quality that is the central fact of consciousness. (Humphrey 1996, 200)

Varela restated cognition not as copy or passive implantation of information into the receiver but as an active form of illumination of the world. (Varela, Thompson, and Rosch 1992) For Varela, knowledge co-evolves with the knower and vice versa. Knowledge implies the other there, co-gnoscere. Varela's metaphor of illuminating with von Uexkull's idea of *Umwelt*, complements the picture of knowledge as a live cognitive process in each organism. The *umwelt* is woven of *presentamina*, objects present to perception and limited by its possibilities of coupling by scale and structure, "locks indifferent to any but their own key" as James writes. Aesthesis is the lamp from inside our body that illuminates a part of the world but blocks all the rest, making us live to the rhythm of a particular neuronal choreography intermittently illuminated. It lights the world through photons in contact with our membranes selecting and differentiating what we require for autopoiesis. At birth, we are born to aesthesis because every birth is opening each creature to the light of the world as well as opening the world to the light of each creature.

SENTINELS OF FORM

There is nothing in the intellect that has not passed, according to Aristotle, through the senses which means, in our terms, that the origin of semiosis and noesis is aesthesis. In *De sensu et sensibilibus* he suggests that feeling is generated in the soul (origin of noesis) through the body (origin of aesthesis) and that the presence or absence of feeling distinguishes an animal from what is not. In *De Anima* he asserts that the powers of souled beings are in ascending order: the nutritive (*threptikon*), appetitive (*orektikon*, linked to the touch), sensitive (*aesthetikon*), locomotor (*kinetikon*), and the power of thought (*dianoetikón*). "Plants have the first while other living things have that one and other powers as well, for instance the aesthetic. If any order of living things has the aesthetic, it must also have the appetitive, as the appetite is the genus where desire, passion, and longing are the species." (Aristotle *De anima* II 3: 414–31) He formulates thus a systemic view of the relative continuity and increasing complexity from plants, animals, and humans. However, for nourishment, nutrients must be selected, and to do so one must perceive them. From our perspective, it is not a line but a cycle in every case, as perception-signification-action are all involved in nutrition. We perceive to signify to value and to act on what is required for sustaining life.

The idea of the evolution of species was already proposed by the presocratic philosophers Anaximander and Empedocles and lay dormant because the logic of the process was unknown. Two millennia later, two great minds asked the same question, independently and simultaneously: Darwin and Wallace answered it with the formula of "random mutation and natural selection" (RM + NS) and finally solved that enigma. However, even today in the supposedly most technologically and scientifically advanced country of the planet, an average of different polls finds that 43 percent of Americans believe that God created humans.[4] The Monkey Trial against John Scopes, a high school teacher indicted in Tennessee during the twenties for having taught the theory of evolution in school, proves this resistance against the evolutionary paradigm among certain social groups like creationists and fundamentalists who to date still maintain a war against it.[5] It seems that the amazing and miraculous fact of evolution itself is not convincing enough, so the idea of God is needed to make it credible. As Antonio Lazcano sums it well: "rendering to Caesar the things that are Caesar's, to God the things that are God's . . . and to Darwin those that are Darwin's."[6]

The novelty of Darwin and Wallace was not the idea of evolution but the explanation of its mechanism. Although Darwin had worked for 20 years before on this concept, both presented to the Linnaean Society in 1858 a paper where they propose that random mutations or changes are incorporated or discarded by genetic drift and natural selection based on performance in the context in which they are located. There may be mutations that do not significantly affect the body, others that benefit it and will be selected by giving better chances of survival and reproduction, and others that can be harmful and undermine these possibilities.

"Linnaeus and Cuvier have been my two gods, though in different ways, but they were mere schoolboys compared to old Aristotle" says Darwin in the first paragraph of *Origin of Species*. He criticizes him for his explanation on tooth formation as an accidental fact and exhibits the affinity and divergence between these two great observers of nature. Quoting Aristotle in his "Physicæ Auscultationes" (lib. 2, cap. 8, s. 2), Darwin writes:

> So what hinders the different parts [of the body] from having this merely accidental relation in nature? As the teeth, for example, grow by necessity, the front ones sharp, adapted for dividing, and the grinders flat, and serviceable for masticating the food; since they were not made for the sake of this, but it was the result of accident. And in like manner as to the other parts in which there appears to exist an adaptation to an end. Wheresoever, therefore, all things together (that is all the parts of one whole) happened like as if they were made for the sake of something, these were preserved, having been appropriately constituted by an internal spontaneity; and whatsoever things were not thus constituted, perished, and still perish. (Darwin 1876, xiii n.)

Darwin adds here: "We here see the principle of natural selection shadowed forth, but how little Aristotle fully comprehended the principle, is shown by his remarks on the formation of the teeth." For Aristotle intentionality determines a priori the features that remain, and for Darwin the case is reversed: survival is the arbiter of subsequent traits.

Aristotle defined life as that which has soul, "actuality of something that has the potential to be souled." The soul would be to the body what the form is to matter or *morphé* to *hylos* and what the actual to the potential. He is perplexed by the fact that "certain plants and animals live even when cut into segments, which means that each segment may have a soul identical in kind, though not numerically identical with different segments, for both segments possess for a time the power of sensation and local movement" and concludes that the principle found in plants is also a kind of soul on its way to actualize matter. (Aristotle *De Anima* I, 5, II 2)

Today Aristotelian hylo-morphism somehow resonates in genes as the sentinels of form (*morphe*) passing it from one generation to the next by devouring matter (*hylos*) into form and replicating it. *Morphé* is constituted by the *hylos* of carbon chains joining amino acids in peptide connections which allow to produce proteins that form coupled tissues in cells associating in organs that structure bodies that constitute organisms in the biomes of the biosphere that conform the great Gaia. All of this is guided by the alert surveillance over form according to the organizing principle of the genes.

CHEMO-AESTHETICS

It is plausible that aesthesis could begin at the cellular level, but to consider that it might arise already from the molecular level is perhaps too risky as it would imply the activity of an agent or some kind of responsiveness at such a rudimentary level. Nevertheless, a suborganismic order selection of proteins to fit antigens in the space of molecular possibilities is proposed by Kauffman in the origin of evolution. Such adaptation requires that the form of the antigen operate as a stimulus or signal on the form that the antibody has to acquire in order to attach.

> But consider autocatalytic polymer systems, perhaps comprising sets of RNA sequences mutually reproducing themselves. Let two such systems interact by exchanging RNA sequences. A sequence injected by the first system into the second might poison that second system such that it no longer is able to reproduce itself. Heritable variation and natural selection—Darwinian evolution . . . might lead to the development in each system of defenses for warding off harmful sequences emitted by the other. Indeed, even the evolution of exchange of those sequences, which are mutually helpful, might occur. In short, autocatalytic polymer systems are primitive examples of a kind of agency, the locus of survival and death, the locus of integrated response to the environment, Given proliferation and Darwinian selection, we have a clear notion of what is "good" for such system, and hence also of the functional

> import of any process to the system's proliferation. Such systems, in the same sense that holds for *E. coli*, come to know their worlds. (Kauffman 1993, 286)

He adds: "It is no small conclusion that heritable variation, and hence adaptive evolution, can occur in a self-reproducing molecular system lacking a genome. Since Darwin's theory of evolution, Mendel's discovery of the 'atoms' of heredity and Weismann's theory of the germ plasm, biologists have argued that evolution requires a genome. False, I claim." (Kauffman 1993, 285)[7]

According to Robert Pollack, what allows transcribing DNA into protein construction is contact (Aristotle's *orektikon*?) of RNA with the figures at the edges of molecules for transcribing them into the amino acids. Thus, the concept exo-endo is relativized as this touching is exo-aesthesic to the ribosome but endo-aesthesic to the cell.

> Proteins and their domains work by recognizing the three dimensional shapes of other molecules. The proteins called enzymes change the molecules that fit into them. The outer surface of an enzyme is indented with pockets of various sizes to fit with remarkable exactitude around another molecule. What happens after the instant of recognition by touch depends on the rest of the enzyme's structure. (Pollack 1994, 73)

If I have not misinterpreted Pollack, he is talking about perception or aesthesis at the level of protein, namely molecular chemo-aesthesis at the edges and grooves of DNA involving semiosis and transcription. He states that the information in DNA sequences is constructed like beads on a rosary by grooves and channels with a definite direction. The two DNA strands run in opposite directions, as the image of the king of hearts in the deck, in a dyadic symmetry that retains its shape after half rotation. Molecules can use the DNA structure for aligning to this transcription of sequences of base pairs in the duplication or replication process.

He argues that cell language is grammatical with a simple syntactic structure reduced to "do this" "to that," or "now, here, translation begins: do this to that." DNA functions as a molecular text processor consisting of 5 letters CATGU and 6 functions: Cut, Paste, Search, Undo, Print, and Copy DNA. Since the statements are discrete, and variable, context-dependent, they do not seem to be the result of a purely mechanical process. As a score, DNA is composed of actions and silences: "Symphony or embryo, the principle is the same: the more complex the pattern, the more important the silences." (Pollack 1994, 73–97, 76)

The case of RNA polymerase, when it detects a molecule of guanine in the DNA strand, cytosine grips molecules from soup swimming in the vicinity and matches the guanine and thus weaves this transcription of molecules until it reaches the "high" according to code instructions. We are referring to actions such as detection, seizing, binding, stopping, or a change of activity involving some kind of agency and the presence of some sort of subject.

> The notion of the polymerase grabbing on to a DNA strand is more than a metaphor. Two relatives of RNA polymerase—DNA polymerase and reverse transcriptase (RT) enzyme from HIV virus—attach to DNA the way a sailor's hand catches a rope: the domains of these proteins both take the precise form of a person's right hand . . . because similar meanings are expressed by similar forms, these two proteins both fold into tiny, hand shaped sculptures. (Pollack 1994, 77–78)

The evolutionary rudiments of sensorimotor processes of multicellular organisms are undoubtedly based on the cellular organisms, analogously as these cellular processes are based on the molecular.

> Messenger RNA like a grooved phonograph record, is a portable representation of the base-pair bumps of its gene, and the translating machinery of the cell, like a juke box, will play the music of any record that brings to it. It will convert the messenger RNA's linear text its sequence of base bumps, into a chain of amino acids. These will then become the meaning of the gene as they fold up into a fully articulate, movable active protein. (Pollack 1994, 80)

Barbieri proposes that the cell is a trinity of genotype, phenotype, and ribotype, and consists of genes, proteins, and ribosoids. (Barbieri 1981, 2008b) For the author, semiosis begins by coding at the ribotype level. His model is based on the fact that a ribotype has historical priority on the genotype and phenotype. For Barbieri, genes and proteins resulted from spontaneous generation in the early Earth—but these proto-structures could not give rise to the cells, as they had no biological specificity. On the contrary, they gave rise to replicating and coding where these molecular machines made of ribosoids became the first cells. (Barbieri 2008b, 27)

There is a lot of speculation today on the possibility that even more rudimentary forms of life or protocells may have been precursors of self-replicating RNA molecules before stabilizing by natural selection in DNA. Such is the case of simple fatty acids (PNA peptide nucleic acid) membranes conforming sacks that keep water, as proposed by Ricardo and Szostak (2009, 38). As with Zeno's arrow, to traverse from inanimate to animate worlds, one needs first to cross half its distance, and half of half, ad infinitum.

CYTO-AESTHETICS

"Thought and behavior in people are rendered far less mysterious when we realize that choice and sensitivity are already exquisitely developed in the microbial cells that became our ancestors." (Margulis 1996, 139) In fact, a shift in focus has occurred that involves taking into consideration an aesthesio-biological approach, which significantly extends the scope of both biology and aesthetics. For Bruni (2008) intercellular communication and signal transduction has caused a paradigm shift in biology into a "science of sensoriality" because feeling or sensation (sensing) is one of the necessary properties of life. The origin of sensitive cells capable of reacting to stimuli and transmitting it to other receptive cells seems to go back to the spirochetes, which led the development of nerve cells in the earliest animal species. If, as proposed by Bruni, metabolic codes and modular properties depend on multimodal signs, cyto-aesthesis and its complement cyto-semiosis keep the clues to the origin and functioning of life. There are "categorical sensations," "redundancy," and "cross-talk" at different hierarchical levels of the cell that can be decrypted by cyto-semiotics.

In this cellular level, communication modes for Prosser include the conductance membrane, patterns of neural spikes, the electrical coupling between cells, electrical and chemical transmission at synapses, neural secretion, and function modification. (Prosser, 1985, 118) The basic principle for understanding neurons boils down to the selective permeability of plasmatic membrane to ions (charged atoms) that appear to penetrate through specific pores or channels in the membrane.[8]

Cyto-aesthesis, as the complement of cyto-semiosis, obviously requires sensory organs or receptors on the cell that allow it to perceive its environment:

> The receptors are molecules . . . and are made up of proteins, tiny amino acids strung together in crumpled chains, looking for something like beaded necklaces that have folded in on themselves. . . . Basically, receptors function as sensing molecules—canners. Just as our eyes, ears, nose, tongue, fingers and skin act as sense organs, so, too, do the receptors, only on a cellular level. (Pert 1999, 23, 22)

Semir Zeki discovers that there are many visual areas of the brain where each receives from the retina specific stimuli directly from area V1, which is the primary visual cortex. (Zeki 2006, 246) He states that the perceptual and processing sites are the same, since cells already select wavelength and project to the area V4, so that vision operates by specialized subcompartments in signal processing. (Zeki 2006, 246, 248–9) Particular areas receive and decode color without having to consult a repertoire of colors stored in memory or central brain. When a node activity has a conscious correlate, Zeki defines it as "essential node." (Zeki 2006, 261) As noted by Logothetis, visual consciousness involves many micro-consciousnesses distributed in space and time, dependent fluctuations of forces between cells of different areas. (Logothetis 1998)

Despite strong confrontation from the scientific community, Lynn Margulis proposed almost half a century ago the revolutionary theory of evolution by symbiogenesis whereby the origin of the eukaryotic cell is in the unicellular organisms by cooperation rather than competition or survival of the fittest from random mutations and adaptive selection. (Sagan 1967) The evidence for this thesis has grown significantly, particularly with the discovery of different types of DNA in a single cell such as mitochondria organelles that we inherit from our mothers. Margulis sharply emphasizes that it is necessary to observe evolution from an infinitely greater time scale than that used by Darwinists who focus only on the last quarter of evolution. We have to go 3000 million years ago, beyond zoology and botany to include chemistry in bacteria, fungi, and protists. We should also observe a much smaller spatial scale inside the cell.

Bacteria are the oldest inhabitants of the planet—kingdom of archaea—origin of life, semiosis, metamorphosis, and more efficient replication. In bacteria, there is sensitivity to the environment from which they feed and adapt with amazing malleability (for instance as regards to antibiotics). They communicate with a molecular language that uses two codes: the exclusive local language and bacterial lingua franca.[9]

A case in point is the *Vibrio fischeri*, a bacterium that lives in seawater and generates a collective effect of luminescence through what is known as quorum sensing. These bacteria are capable of distinguishing states when they are relatively isolated from those in which they are accompanied. When they perceive a critical mass of similar bacteria in the environment at relative proximity, they secrete molecules that make them illuminate synchronously in collective chemical communication.

Evolution is essentially a process of communication from the eukaryotic ribosomes or nucleated cells (originally symbiotic, and as such necessarily also semiotic) implying an evolutionary leap from intracellular semiosis to intercellular—albeit at the cost, according Sonea, of giving up open global bacterial communication. (Sonea 1992, 385–6) Different communication processes between bacteria depend on ongoing exchanges and unmediated proximity between numerous agencies. In contrast, the eukaryotic cell closes its membraned core of information for selectively establishing exchanges between the exterior and interior.

Pharmacists believed that if drugs act in the body there had to be some element in it to which they adhered, as Paul Ehrlich points out in the early twentieth century. No drug can act unless it is fixed within the body. The term "receptor" was then used to refer to this hypothetical component that enabled latching-on allowing the drug to initiate the cascade of physiological changes. "Now we know that that component, the receptor, is a single molecule, perhaps the most elegant, rare and complicated kind of molecule there is," writes Pert (1999, 21) In 1973, Candace Pert pioneered investigations on opioid receptors. She describes the cellular processes of perception: "A typical neuron (nerve cell) may have millions of receptors on its surface. Molecular biologists can isolate these receptors, determine their molecular weight, and eventually crack their chemical structure, which means identifying the exact sequence of amino acids that makes up the receptor molecule." Drugs can be designed specifically for these receptors.

> They hover in the membranes of your cells, dancing and vibrating, waiting to pick up messages carried by other vibrating little creatures, also made out of amino acids, which come cruising along—*diffusing* is the technical word—through fluids surrounding the cell. We like to describe these receptors as "keyholes" although that is not altogether precise term for something that is constantly moving, dancing in a rhythmic, vibratory way. (Pert 1999, 23)

> They wiggle, shimmy, and even hum as they bend and change from one shape to another, often moving back and forth between two or three favored shapes or conformations. In the organism they are always found attached to a cell, floating on the cell surface's oily outer boundary or membrane. Think of them as lily pads, floating on the surface of a pond and, like lilies, receptors have roots enmeshed in the fluid membrane snaking back and forth across it several times and reaching deep into the interior of the cell. (Pert 1999, 22)

Pert quotes Miles's findings that cells have specialized receptors for light, sound or touch as adaptations that pick up signals on the tasks to be done for keeping their balance. In fact, apparently less than 2 percent of communication actually occurs at the synapse, so it follows that the brain is organized not as much by synaptic connections as by the specificity of the receptors in binding to one type of ligand. (Pert 1999, 139)

She adds that "all receptors are proteins. . . . And they cluster in the cellular membrane waiting for the right chemical keys to swim up to them through the extracellular fluid and mount them by fitting into their keyholes, a process known as *binding*. Binding. It's sex on a molecular level." (Pert 1999, 23) Mobile cells of the immune system called macrophages (or pathogens and debris-eating cells) detect pathogens and toxins and cleanse the blood to purify the body. "[A]ntibodies vibrated and changed as they encountered the bacteria, virus, cancer cells, *latching on* to them and escorting them out of the system." (Pert 1999, 160)[10] Such macrophages exploit bacterias' perceptual biases by producing antibodies by the antigens' shapes to "seduce" them and latch them out. Proteins are messages or information vehicles to activate or paralyze cells. Certain proteins called interferons are produced by various cells to inhibit viruses' replication by activating macrophages and lymphocytes or white blood cells that reproduce to stop viral invasion.

In general, when speaking of receptors in the organism, we think of cell membrane parts that are coupled to a drug or a virus molecule.

> In the wake of discoveries in the 1980s, these receptors and their ligands have come to be seen as "information molecules"—the basic units of a language used by cells throughout the organism to communicate across systems such as the endocrine, neurological, gastro-intestinal and even the immune system. Overall, the musical hum of the receptors as they bind to their many ligands, often in far-flung parts of the organism, creates an integration of structure and function that allows the organism to run smoothly, intelligently. (Pert 1999, 27)

What makes endosemiosis or communication so efficient within each organism is the fact that information is issued in an organized way according to the specialized conformation of each receptor. At the cellular level, Patee proposes that there is a semiotic closure for the emergence of self, where the given semantic description and materials construction dynamics can operate in the interplay of strong and weak ties. He wonders about the central problem of evolution and asks what types of symbolic descriptions, control restrictions, and material constructions promote the survival of populations?[11] That a cell can "sniff" the presence of a peptide by chemotaxis and follow it until latching on to it with its receptor allows us to conjecture that similarities across different scales in the three levels, cellular, multicellular, and multi-individual, may be greater than we think.

Cell Emotionality

Neither is the thrill in the heart nor the mind in the head. All creatures, regardless of scale or complexity, have bodily processes, mental and emotional tones that operate together to preserve our lives. So not only can the *Escherichia coli* know their world, but feel it and appreciate it, even as emotional suffering that results from corporeality itself. Our body is always emotional. James proposes a radical transformation in the way we understand emotion, not as an elusive effluvium emanating from the heart or mind of a human being but as bodily responses to the environment, concrete organic changes in the body such as muscle tension, sweaty palms, nausea, agitation. Consequently, every corporeal creature is emotional, whether its body is cytoplasm or an elephant, although the way it feels for such creature, the qualia, is only accessible to such creature.

> What kind of an emotion of fear would be left, if the feelings neither of quickened heart-beats nor of shallow breathing, neither of trembling lips nor of weakened limbs, neither of goose-flesh nor of visceral stirrings, were present, it is quite impossible to think. Can one fancy the state of rage and picture no ebullition of it in the chest, no flushing of the face, no dilatation of the nostrils,

no clenching of the teeth, no impulse to vigorous action, but in their stead limp muscles, calm breathing, and a placid face? The present writer, for one, certainly cannot. The rage is as completely evaporated as the sensation of its so-called manifestations, and the only thing that can possibly be supposed to take its place is some cold-blooded and dispassionate judicial sentence, confined entirely to the intellectual realm, to the effect that a certain person or persons merit chastisement for their sins. In like manner of grief: what would it be without its tears, its sobs, its suffocation of the heart, its pang in the breast-bone?

A feelingless cognition that certain circumstances are deplorable, and nothing more. Every passion in turn tells the same story. A purely disembodied human emotion is a nonentity. I do not say that it is a contradiction in the nature of things, or that pure spirits are necessarily condemned to cold intellectual lives; but I say that for *us*, emotion dissociated from all bodily feeling is inconceivable. The more closely I scrutinise my states, the more persuaded I become, that whatever moods, affections, and passions I have, are in very truth constituted by, and made up of, those bodily changes we ordinarily call their expression or consequence; and the more it seems to me that if I were to become corporeally anaesthetic, I should be excluded from the life of the affections, harsh and tender alike, and drag out an existence of merely cognitive or intellectual form. Such an existence, although it seems to have been the ideal of ancient sages, is too apathetic to be keenly sought after by those born after the revival of the worship of sensibility, a few generations ago. (James 1884)

James's intuition is confirmed a century later by the discovery of peptide polymers, messengers whose receptors are located in the nervous system as well as in the endocrine and immune systems through a peptidergic communication web throughout the body. (Pert 1999, 285) Pert finds that "where the peptides and their receptors are richest are also the parts of the brain that have been implicated in the expression of emotion . . . the limbic system had the densest concentration of these receptors." (Pert 1999, 178) She understands that emotions are chemical substances as ligands or peptides circulating through the intercellular fluid to which specific cell receptors in the brain and the whole body latch on and generate effects in mood, behavior, and memory. In an eloquent metaphor she affirms that: "Peptides are the sheet music containing the notes, phrases and rhythms that allow the orchestra—your body—to play as an integrated entity. And the music that results is the tone or feeling that you experience subjectively as your emotions." (Pert 1999, 148) Emotion is the tone in the sense of being there.

Emotional states or moods are produced by the various neuropeptide ligands, and what we experience as an emotion or a feeling is also a mechanism for activating a particular neuronal circuit—*simultaneously throughout the brain and body*—which generates a behavior involving the whole creature, with all

> the necessary physiological changes that behavior would require. (Pert 1999, 145)

Emotion is the effect of intercellular communication, a molecular chemical process initiated in cells that can trigger further reactions. Sending signals to produce molecular actions between systems confirms that emotion is not just an individual process, but also a refined tonal, almost symphonic or cacophonic resonance to another (cell or organism) through bodily and cultural indexes that, in the case of culture, can moreover be evoked by means of literature, painting, music, theater, as well as by face to face interaction.

The Fascinating Case of the Loving Yeast

Yeast is unwilling to marry just anyone even if she is such an elementary unicellular organism, because in that marriage she bets it all out. The critical decision to mate or not to mate with a certain prospect would be based not on a particular pheromone but in the contact between the propheromone on yeast's cell membrane and the receptors in the membrane of another cell. Therefore, it appears that yeast cells actually select whom to mate with, as they "like" and "prefer" to mate with those who secrete more pheromones. According to Edna Nahon:

> Yeasts are unicellular organisms. Yeasts of the species *Saccharomyces cerevisiae* (baker's yeast) have two genders, and alpha and a, that mate with one another. Each of them secretes a particular peptide and has a special receptor for peptide secreted by the other. Detecting the peptide of the opposite gender leads the individual receiving it to stop growing and begin processes that prepare it for mating. Such mating involves fusion of the two cells to form a body. Mating involves merging the two cells together to form one organism. Mating in yeasts, as in other unicellular organisms, is an all or nothing proposition: each has only one chance to mate in its life. [12]

Even at this primal level, yeast is able to select a mate rather than coupling randomly because "cells can advertise their phenotypic quality by means of peptides and their propeptide progenitors, which hold them in specific spatial configurations." (Zahavi and Zahavi 1997, 104) To announce quality requires an appraiser sensitive enough to appreciate it. Such is the role of the alpha propheromone anchored in the cell membrane by its hydrophobid segment to protrude sections outside of the membrane in a particular spatial configuration. "The pheromone molecules reach the other gender, are sensed by its receptors, and trigger the receiving cell into sending a copulating offshoot toward the cell secreting the pheromone." (Zahavi and Zahavi 1997, 106)

And when the yeast found her prince charming they married and lived happily ever after eating sugars, salts, and vitamins and using oxygen to burn food. By losing nutrients and suitable temperature, the yeast cancels the process of transcription and translation of genes, activates a few and closes itself within a thick spore wall, hibernating. That way it can remain quiet unless it is invaded by a virus that latches to her body and exploits it with instructions to replicate the invader. In this viral latching, a reading error occurs despite the high specificity of their receptors that is lethal for the cell.

This leads us to wonder whether fertilization is a purely electrochemical or mechanical operation (the faster or more vigorous sperm arrives first to the ovule, as in Woody Allen's film) or a deliberate choice process of chromosomes in the sperm by the ovule already guiding the evolution of the species from this cellular level. It is possible that, as in the chimpanzee, where natural selection takes place within the uterus, women's sexual selection that so dazzled Darwin in the *Descent of Man* already begins at this microscopic scale. Moreover, there have been sperm couples joined at the head under a common hat who help each other swim together to accelerate and fertilize the egg.[13] A sensible ovule, thinking in her progeny, would definitely choose a cooperative sperm. Many surprises await us in the field of cyto-aesthesis.

PHYTO-AESTHETICS

A more careful observer even than Aristotle, and certainly more modest because he only explores nature without attempting to cover also the ethics, politics, metaphysics, rhetoric, and poetics, Darwin contradicted his teacher when he claimed that plants do have movement. He remarks that:

> It has often been vaguely asserted that plants are distinguished from animals by not having the power of movement. It should rather be said that plants acquire and display this power only when it is of some advantage to them; but that this is of comparatively rare occurrence, as they are affixed to the ground, and food is brought to them by the wind and rain. We see how high in the scale of organization a plant may rise, when we look at one of the more perfect tendril-bearers. (Darwin 1865, 118)

After studying the vegetation by geographical areas and processes of fertilization and hybridization, he published in 1865 that "[t]he most interesting point in the natural history of climbing plants is their diverse powers of movement; and this led me on to their study. The most different organs—the stem, flower-peduncle, petiole, mid-ribs of the leaf or leaflets, and apparently aërial roots—all possess this power." (Darwin 1865, 115.) He questions Hugo von Mohl's explanation that the rotation of the plant's axis causes the revolving movement: "But the best evidence that the twisting does not cause the revolving movement is afforded by many leaf-climbing and tendril-bearing plants . . . of which the internodes are not regularly twisted, but which regularly perform, as we shall hereafter see, revolving movements like those of true twining-plants." (Darwin 1865, 5)[14]

Darwin examined more than 100 plant species and what prompted his curiosity initially, as he himself declares, was an article by Asa Gray on the "extreme sensitiveness and rapidity of movements in the tendrils of certain Cucurbitaceous plants." (Darwin 1865, note in 49) The first thing he discovered is the sensitivity of the petioles, and the manner in which the stems of flowers as *Maurandia* are tender to touch. He notes that there are plants that have a hook at the end that allows them to find where to stop and avoid being detached by windstorms. (Darwin 1865, 8)[15] On the *Clematis Montana*, he writes the following:

> [T]he long and thin petioles of the leaves, whilst young, are sensitive, and when lightly rubbed bend to the rubbed side, subsequently becoming straight. They are far more sensitive than the petioles of C. *glandulosa*; for a loop of thread weighing a quarter of a grain caused them to bend; a loop weighing only one-eighth of a grain sometimes acted and sometimes did not act. The sensitiveness extends to the angle between the stem and leaf-stalk. (Darwin 1865, 27)

At first, Darwin attempts to refute Henslow's position for whom the spiral growth of plants is a mechanical fact. He starts from here a debate that has not been resolved to this day in the struggle between a mechanistic and deterministic approach versus another that assumes some kind of agency, greater flexibility and choice. Darwin opts for the latter. It is ironic that despite his position on this, Darwin has been used to support mechanistic positions and accused of biological determinism (i.e., social Spencerianism).

> Many authors speak as if the movement of the plant to light were a direct result of evaporation and oxygenation of the sap in the stem, as in the expansion of an iron bar by increasing its temperature. But seeing that the tendrils are neither attracted nor repelled by light, it is more likely that their movements are guided and encouraged only by their action, in the same way as they are led by the attractive force from or towards the center of gravity. (Darwin 1865, 116)

The first case tries to explain a phenomenon of *Orbis secundus* by concepts that belong to *primus* such as evaporation and oxygenation regardless of the difference or qualitative changes in scale (contradicting Rothschild's third biosemiotic law on hierarchic organization dealt with in the previous chapter). Darwin, by contrast, assumes that difference and highlights aspects that influence the *Orbis primus* in plant morphology but not determine it. The aforementioned distinction between *motion* (which is inertial in *primus*) and *action* (which depends on agency or intention in *secundus*) summarizes Darwin's position when implying that there is action, not just motion, in plants.

> We see how high in the scale of organization a plant may rise when we look at one of the more perfect tendril-bearers. It first places its tendrils ready for action, as a polypus places its tentacula. If the tendril be displaced, it is acted on by the force of gravity and rights itself. It is acted on by the light, and bends towards or from it, or disregards it, whichever may be most advantageous. During several days the tendril or internodes, or both, spontaneously revolve with a steady motion. The tendril strikes some object, and quickly curls round and firmly grasps it. In the course of some hours it contracts into a spire, dragging up the stem, and forming an excellent spring. All movements now cease. By growth, the tissues soon become wonderfully strong and durable. The tendril has done its work, and done it in an admirable manner. (Darwin 1865, 118)

The coiling about an axis as in the rope has the function of providing more rigidity to the stem. Darwin marks the stems with stripes and observes in which direction they move, verifying that one turn is concave and the next convex measuring revolutions in hours and minutes. He notes that most turn towards the sun's movement and only some species do the opposite. All authors reviewed on growth of vines assume it is due to the nature of the plant, except Mohl, who mentions some irritability (which translates to sensitivity "but the sensitiveness or irritability"). (Darwin 1865, n 26)

Darwin lightly rubs tendrils or vines sprigs that extend through a surface to latch onto it and finds no effect. He concludes that curved stems are not irritable, and it is not likely they would be so, since nature always economizes its means and such irritability would be superfluous. Immediately after stating that, he retracts and can not deny some degree of irritability:

> Hence I conclude that twining stems are not irritable; and indeed it is not probable that they should be so, as nature always economizes her means, and irritability would be superfluous. Nevertheless I do not wish to assert that they are never irritable; for the growing axis of the leaf-climbing, but not spirally twining, *Lophospermum scanden*s is, as we shall hereafter see, certainly irritable; but this case gives me confidence that ordinary twiners do not possess this quality. (Darwin 1865, 10)

The subject interested him so much that he persists in this research with his son Francis, and in 1880 published the book *The Power of Movement in Plants*. The Darwins diligently placed cardboards and wood chips to measure "circumnutation" (helical rotary motion in growing organs) to find the deviation when reaching obstacles in the soil or detecting irritants. They ascertain plant sensitivity to touch, light, and temperature even on germinal seed that develops in the first leaves.

Plants latch on to the ground and can be fixed but not still. They emerge through trees' foliage competing for light as lianas, sometimes taking unpredictable forms, in contrast to animals whose anatomy is completely predetermined by their species. Phyllotaxis is the organization of the leaves around the stem to achieve greater exposure to the mist and sun. Thus the shape of a plant is determined not only by genes but by the environment since it has an enormous plasticity, such as meristems, allowing it to accept grafts of different types (or what is known as genomic fluidity that animals lack and is replaced by behavioral flexibility).[16] They pay for their relative immobility with their ability to capture sunlight for photosynthesis, the principle on which all life on earth depends. Half of the ground biomass is constituted by the same type of carbohydrate, cellulose, a molecule that stiffens the cell walls forming the vegetable world and 99 percent of multicellular organisms. (King 1997, 165)[17]

If perception can be measured by action (as James measures emotions by corporeal reactions) we can say that a plant sensor or phyto-preceptor exists that is capable of detecting space, surface, nutrients, water, and light and manifests it as a phyto effector by the action performed towards such objects. Plants act because they have sensitivity confirming the triplet aesthesis-semiosis-praxis also at this phyto-semiotic level.[18] This is another way of referring to von Uexküll's functional cycle which we translate as a process of receptor (aesthesis) that encodes/interprets (semiosis) and acts as an effector (praxis) (although perceiving and interpreting are always already actions, so the term praxis is implied and could be considered redundant).

Plant Sensoriality

One of the key elements for the survival of plants is their sensitivity to CO_2. The plant world requires special sensors for these molecules, but little is known how sensitivity to this gas occurs. It is also unclear how plants detect sugar or calcium, but an intracellular mechanism has been found by which a plant hypersensitive to sodium, signals calcium in mediating tolerance to salt.[19] Another aspect vital to the survival of the plant is its sensitivity to gravity, particularly in the early stages of seed germination. The aesthesis to gravity is crucial to guide the movement towards light and root to gravity, which is termed as gravitropic response. This response is manifested particularly by its formation through growth that is woven between increased exposure to light and its balance in relation to gravity. (Hangarter 2008) Theodore de Saussure discovered that plants are even able to select from minerals taken from the earth.[20]

The sensitivity of various plants to light is remarkable in *Albizia saman* that folds its leaves at night and on cloudy days or in *Gentiana nivalis* flowers that close their petals even when a cloud is passing through the sun's light. Heliotrope plants' rotating towards the sun is notable to maximize their exposure to photons required for photosynthesis and energy for their metabolism.

Phytochrome is the light receptor essential for guiding seed germination, growth, flower formation, and adjusting the circadian rhythms. It is located from the stem end to the roots for sending different signals depending on its position. (King 1997, 112–13) Since its discovery in 1960, it is known that phytochrome is a blue pigment that absorbs red and infrared light and regulates that some leaves and petals bend at dark. In addition to phytochrome, phototropins and cryptochromes are light receptors existing in plants.

Effects of plants' sensitivity are outstanding in species like *Clematis calycina*, as noted by Darwin, because when comparing two stalks, one in contact with a surface and the other not, the former almost doubles the thickness. The contrast in cell size of each case can be seen through the microscope, but the most striking difference is the hardness of the surface as the stems adhered to a surface become hard and difficult to cut, while the others remain soft.

In *Clematis viticella*, Darwin studied various degrees of sensitivity according to each part of the plant: the medial and lateral petioles are somewhat more sensitive, the basal portion is less responsive although they react to a contact surface. Finally the lower surface of the terminal portion forming the inner side of the hook is the most sensitive, so it can attach itself to possible distant supports. To test the sensitivity of the plant, Darwin placed wire hoops in the lateral and terminal subpecioles; in a few hours the ends were folded while the laterals showed no effect twenty-four hours later. (See quote below)

Plants feel the presence of a rod and by their position, grow around it or over it. He notes that *Clematis flammula* is the most sensitive of all the plants he knows, but there are clear differences between the areas and degree of maturity of each part. It also depends on the time of year as spring seems to be the period of greatest sensitivity in this species. (Darwin 1865, 32) The speed of the response of a plant to actions upon it, indicates their degree of sensitivity.

> A thin stick placed so as to press lightly against a petiole, bearing a leaflet a quarter of an inch in length, caused the petiole to bend in 3 h. 15 m.; in another case a petiole curled completely round a stick in 12 h. These petioles were left curled for 24 h., and then the sticks were removed; but they never straightened themselves. I took a twig, thinner than the petiole itself, and lightly rubbed with it several petioles four times up and down; these in 1 h. 45 m. became slightly curled; the curvature increased during some hours and then began to

> decrease, but after 25 h. from the time of rubbing a vestige of the curvature remained. (Darwin 1865, 32–3)

The *Trillium grandiflorum* showed to be very sensitive to touch, because the slightest friction causes them to bend to the side rubbed in about three minutes. (Darwin 1865, 359) One bent into a ring in six minutes then straightened again. However, once they have fully attached to a stick, when removed, they do not return to their original form.

Different species may react similarly or not in different times and with various degrees of sensitivity. In *Tropaeolum peregrinum*, Darwin noted the degrees of sensitivity in relation to the weight of the wire rings. He contrasts *Tropaeolum tricolorum* and *T. tuberosum*, the first "exquisitely sensitive" and the second practically insensitive or quite atrophied that faculty or lost it or never acquired that sort of sensitivity. Sensitivity may vary depending on the maturity and position in the tendrils and even in medial, lateral, or terminal petioles. When younger they are often more sensitive than when older. Sensitivity also exists in the leaves, particularly the tip of the leaves, which can be attached to a stick as *Adlumia cirrhosa*.

> The hooked tips of the leaves of the *Gloriosa* are sensitive only on their inner or inferior surface. The petioles are sensitive to a touch and to excessively slight continued pressure, even from a loop of soft thread weighing only the one-sixteenth of a grain; and there is reason to believe that the rather thick and stiff petioles of *Clematis flammula* are sensitive to even a less weight when spread over a wider surface. (Darwin 1865, 47)

In this case, sensitivity depends on coiling, because if the loop is closed it loses sensitivity. In this same species mature and flourishing plants, that would not have grown taller, had leaves without sensitive tips and could not surround a stick. The tip sensitivity in this plant was not a necessary function for survival. "Hence we see that the sensitiveness of tendrils is a special and localized capacity, quite independent of the power of spontaneously moving; for the curling of the terminal portion from a touch does not in the least interrupt the spontaneous revolving movement of the lower part. In *Bignonia unguis* and its close allies the petioles of the leaves, as well as the tendrils, are sensitive to a touch." (Darwin 1865, 101)

Finally, Darwin concludes about his tests on plant movement and discernment that their sensitivity and action in their receptors on the tips of the roots is different from that in their leaves and tendrils.

The Actress and Her Mocked Lover

We cannot conclude this section without mentioning the amazing, almost operatic staging of *Ophrys*, the orchid flower that resembles a female wasp to the male hornet. This flower has similar color, shape, texture, costumes, lighting, pose, props, and even smell in a perfect mimicry of the villus pheromone, identical to the wasp queen:

> The brilliant polished surface of the center of the flower shines like a blue-steel looking glass, edged with gold and set in a velvety maroon background. . . . The violet-blue center of the flower resembles the reflections from the halfway-crossed wings of a resting female. A thick set of long, red hairs imitates the hair found on the insect's abdomen. The antennae of the female wasp are beautifully reproduced by the upper petals of the orchid which are dark and threadlike. (Trivers 1985, 401–3)

This seduction process operates by three steps in three registers: The somatic smell to permeate the air with pheromones like a female's molecules, the scopic to simulate the visual appearance of the female body, and finally guided by the somatic to entice and touch the male in this Capilo-topography of vellosity through the right direction of copulation, who is so excited that he penetrates deep enough to soak pollen and propagate it. By dispersing the orchid's gametes, the male wasp wasted his, because intercourse with the flower lasted for 20 minutes. Plant intelligence in this case exceeds animal intelligence, and the *Ophrys* deserves at least the Academy Award for Best Actress in the kingdom of plants.

Another interesting case is that of mimicry in plants that live in proximity to others and start looking alike by the transfer of genetic information, such as *ariocarpus agavoides* imitating agave. (Hallé 2002, 169, 232–3) Plants such as yams or sweet *Ipomoea batatas* recognize their own and others' pollen and do not fertilize themselves with their own by self-incompatibility. The tricks of nature are such that plants as *passiflora candollei* produce yellow shapes in their leaves mimicking eggs to fool butterflies and foil them from depositing theirs and so they thwart caterpillars to devour their leaves. (Trivers 1985, 404 figure 16.9).

As Hallé noted, plants try to increase their linear and surface dimensions in detriment of their volumes, as they are dependent on absorbing photon energy from the sun. As the solar energy reaches the body from multiple directions, proper placement ensures their better surface to capture it. Plants are essentially exchange surfaces, while animals are mobile to minimize volume surfaces and therefore energy loss, exposure to predators, and limited mobility.

The aesthetics of the plants is displayed mainly for seducing the animal kingdom (bees, hummingbirds, butterflies, and bats) by offering perfumes, colors, and shapes as well as food. It even drugs them with hallucinogens as in *Datura innoxia*, in exchange for transporting pollen and facilitating reproduction. Bees manifest an aesthetic bias for their attraction to red, white, or blue petals; birds prefer orange and yellow, and hummingbirds are fond of red and purple flowers as butterflies to pink or violet. The shapes and designs of flowers, especially their pigments such as carotenoids and anthocyanins, guide the pollinator to nectar. (Hallé 2002, 162–3; King 1997, 147) That is why they hide their seeds with the green color in immature and odorless, hermetically sealed fruit in a hard and often toxic core and thorns.

Apart from attraction tactics, plants also deploy repulsion strategies through thorns, thistles, nettles, hairy textures with hard points or irritants that scare off predators. Another way of repulsion is the phenomenon of *allelopathy* to prevent the growth of other plants in a radius of twenty-five meters, as in the walnut trees (*Juglans regia* and *nigra*). Walnut trees (*Juglans regia)* suppress the growth of neighboring plants within a radius of several meters, as much as twenty-five meters for a large tree. The leaves and branches produce a toxin, juglone, which is leached by rain, enters the soil, and blocks the germination of seeds, giving the walnut exclusive use of the site's resources. (Hallé 2002, 156–7)

Hallé mentions that the acacias produce an astringent taste, by the tannin, which repels the kudus after a few minutes, and so alert the rest of the acacias by the wind to become astringent without being attacked. (Hallé 2002, 157; King 1997, 179)

The semiotics and aesthetics of plants are so complex that we do not even understand how, for example, a plant invaded in its leaves begins to mobilize defenses from its roots or other parts, such as tobacco producing nicotine or tomato producing jasmonic acid to prevent the predator's digestion and when attacked secrete phytoalexins, a defensive antibiotic. (King 1997, 174) Phyto-aesthesis detects different gases and chemicals in the air, temperature, wind, gravity, nutrients, moisture, and light, proximity of other bodies, toxins, and irritants such as caustic. In conclusion, plants are everything but vegetative in a pejorative sense: they deploy strategies, are active and exhibit great sensitivity and wit. Plants also have feelings, as a wounded acacia and the apple who mourns the loss of her freshness by crying ethylene.

ZOO-AESTHETICS: A NATURAL STEP AFTER DARWIN[21]

As Aristotle was to poetics, so was Darwin to zoo-aesthetics. The first essay on animal aesthetics was published in 1871 and, without any feelings of philosophical inferiority, Darwin entitled it "The sense of beauty." And as tragedy was the paradigmatic example for poetics, the aesthetic behavior of peacocks and bowerbirds was equally so for zoo-aesthetics. I transcribe verbatim Darwin's canonical text on aesthetic behaviors observed in various animal species.

> Sense of Beauty. This sense has been declared to be peculiar to man. I refer here only to the pleasure given by certain colours, forms, and sounds, and

which may fairly be called a sense of the beautiful; with cultivated men such sensations are, however, intimately associated with complex ideas and trains of thought. When we behold a male bird elaborately displaying his graceful plumes or splendid colours before the female, whilst other birds, not thus decorated, make no such display, it is impossible to doubt that she admires the beauty of her male partner. As women everywhere deck themselves with these plumes, the beauty of such ornaments cannot be disputed.

As we shall see later, the nests of humming-birds, and the playing passages of bower-birds are tastefully ornamented with gaily-coloured objects; and this shews that they must receive some kind of pleasure from the sight of such things. With the great majority of animals, however, the taste for the beautiful is confined, as far as we can judge, to the attractions of the opposite sex. The sweet strains poured forth by many male birds during the season of love, are certainly admired by the females, of which fact evidence will hereafter be given. If female birds had been incapable of appreciating the beautiful colours, the ornaments, and voices of their male partners, all the labour and anxiety exhibited by the latter in displaying their charms before the females would have been thrown away; and this it is impossible to admit.

Why certain bright colours should excite pleasure cannot, I presume, be explained, any more than why certain flavours and scents are agreeable; but habit has something to do with the result, for that which is at first unpleasant to our senses, ultimately becomes pleasant, and habits are inherited. With respect to sounds, Helmholtz has explained to a certain extent on physiological principles, why harmonies and certain cadences are agreeable. But besides this, sounds frequently recurring at irregular intervals are highly disagreeable, as every one will admit who has listened at night to the irregular flapping of a rope on board ship. The same principle seems to come into play with vision, as the eye prefers symmetry or figures with some regular recurrence. Patterns of this kind are employed by even the lowest savages as ornaments; and they have been developed through sexual selection for the adornment of some male animals. Whether we can or not give any reason for the pleasure thus derived from vision and hearing, yet man and many of the lower animals are alike pleased by the same colours, graceful shading and forms, and the same sounds.

The taste for the beautiful, at least as far as female beauty is concerned, is not of a special nature in the human mind; for it differs widely in the different races of man, and is not quite the same even in the different nations of the same race. Judging from the hideous ornaments, and the equally hideous music admired by most savages, it might be urged that their aesthetic faculty was not so highly developed as in certain animals, for instance, as in birds. Obviously, no animal would be capable of admiring such scenes as the heavens at night, a beautiful landscape, or refined music; but such high tastes are acquired through culture, and depend on complex associations; they are not enjoyed by barbarians or by uneducated persons.

Many of the faculties, which have been of inestimable service to man for his progressive advancement, such as the powers of the imagination, wonder, curiosity, an undefined sense of beauty, a tendency to imitation, and the love of excitement or novelty, could hardly fail to lead to capricious changes of customs and fashions. I have alluded to this point, because a recent writer[73] has oddly fixed on Caprice "as one of the most remarkable and typical differences

> between savages and brutes." But not only can we partially understand how it is that man is from various conflicting influences rendered capricious, but that the lower animals are, as we shall hereafter see, likewise capricious in their affections, aversions, and sense of beauty. There is also reason to suspect that they love novelty, for it own sake. (Darwin 1882, 92–93)

Since aesthetics has been considered the peak of human refinement, this odd idea of animals having a sense of aesthetics was questioned by diverse authors, beginning with Darwin's own colleague in the evolutionary paradigm, Alfred R. Wallace, who criticized Darwin's attribution of sophisticated human emotions to supposedly lower creatures. Wallace denies that aesthetic criteria have any role in sexual selection, and explains dimorphic coloration, which distinguishes male from female in certain bird species, by applying two concepts: 1) protection, since the brown colors of the females camouflage and protect them, and 2) recognition, to distinguish individuals of the same species in breeding and preservation. (Cronin 1993, 157–8)[22] However, as Romanes points out, while this may account for the females' brown color, it does not explain the males' bright colors, which by the principle of economy in nature would tend to be brown. Wallace takes an unexpected turn to explain beauty betraying the evolutionary paradigm by claiming that art and morality are the result of the intervention of a divine creator during human evolution. (Wallace 1867; 1889) Darwin, on the contrary, insists that the explanation of how males acquired such bright colors is sexual selection, and to the present day, we have not found a more convincing account.

When considering animal aesthetics, two meanings of the term are implied: 1) the beauty of animals to human appreciation, and 2) animals' own preferences according to pattern or form. The first was enthusiastically developed by Adolf Portman (1967) who teaches contemplation of beauty in animal features, such as feather patterns and beaks, much like an art critic praising nature's artworks.

As for the second, aesthetics of the animal's own appreciation, and its consequences, there is much evidence in Darwin and many others' observations. Let me only mention two cases to begin with. In a brief essay published five years before Darwin's, James Shaw (1866) highlights in "Feeling of Beauty Among Animals," the astonishing behavior of passerine birds that build nests exclusively for love, shaping arches and decorating them with shells, shiny objects, feathers, glass, and ceramic pieces. Snowdon and Teie's (2010) experiment with tamarins concluded that these animals are capable of reacting to music composed with their own appeasing or alarm vocalizations put together in new, musical sequences. They found that monkeys can be emotionally sensitive to such patterns of vocalization, if not to the quality of the music itself.[23]

As if it were not already obvious, in this book I side with Darwin's stance on animal aesthetics. There is already plenty of empirical support showing how selective different species are, based on criteria of color, proportion, poise, and dexterity. Hence I will focus on the theoretical underpinning of this position as a direct consequence of a biosemiotic and evolutionary approach.

> To grasp the sources of esthetic experience it is, therefore, necessary to have recourse to animal life below the human scale. The activities of the fox, the dog, and the thrush may at least stand as reminders and symbols of that unity of experience which we so fractionize when work is labor, and thought withdraws us from the world. The live animal is fully present, all there, in all of its actions: in its wary glances, its sharp sniffings, its abrupt cocking of ears. (Dewey 1980, 18–19)

Every creature is an affective emotional being because, as William James points out, emotion is the way the body reacts to environmental stimuli. If emotions are chemical reactions, we can consider that even a cell may undergo some kind of emotion at its own chemical level. Darwin insists that animals are not machines but creatures sensitive to a variety of environmental aspects and behaviors in individuals of their own as well as of other species. In his book *Expression of the Emotions in Man and Animals*, he looks for innate universal gestures and distinguishes between learned displays of emotions in many species, even among invertebrates who may feel misery and joy, pleasure and pain. Various vertebrates, including insects, have opioid receptors, a fact that implies the possibility of the feeling of pleasure in these species.

In addition, several animals display admirable gestures of altruism, as the case of Tursiops dolphins registered in a desperate attempt to raise a newborn above the water to try to help him breathe and save his life. Others have altruistic gestures as the case of slime mold assembling when food is scarce to send offspring in spores to other areas, or the adoption of orphans by wolves or chacma papios, the expression of affection between ants by rubbing their antennas and the consolation song of crickets to solitary females. (Hamilton, Busse, and Smith 1982) Cobras show affection for each other. Among budgies it often happens that when one dies, the other cries and long laments its death. Spiders have great affection for their eggs that they carry covered in a silk shawl. Birds adopt orphans or abandoned chicks even of other species and care for adults who have been blinded; curiously they even express deep antipathy against certain individuals without apparent cause. This fine discrimination ability becomes so powerful that Darwin mentions the case of a certain female peacock who not being able to mate with the male of her choice, preferred to remain single throughout the whole breeding season rather than mating with another male. In humans, this behavior is usually far less selective and certainly much less supportive than the male beetle *Ateuchus* who stridulates to encourage the female in her daily chores. (Darwin 1882, 306)

The art of perfumes has its very sophisticated roots in the animal kingdom. In common with cyto-aesthetics and probably phyto-aesthetics too, we can consider the deployment of aromatic molecules as a form of corporal aesthesis to seduce the opposite sex by crocodiles, mammals, and other species or to ward off predators (as *heliconidæ* butterfly). The female moth emits a powerful pheromone to seduce the male. (Wilson 1980, 94) Darwin notes that the aromatic deployment of males is a form of peaceful competition that excites and entices females into mating. Males are usually ready to mate with almost any female and try even harder to seduce the female than to compete with another male, which speaks of the power of female choice. (Darwin 1882, 227) In the mammalian somatic expression, perhaps there is no species that does not emit aromatic molecules to attract or that has no receptivity to detect them. Cases-in-point are the deer (*Moschus moschiferus*), some ruminants (*Bos moschatus*), goats, antelopes, sheep, bats (*Cheiroptera*), monkeys, and, of course, cats and humans.[24] The black-tailed deer *Odocoileus hemionus* produces seven different pheromones in different places and for different functions: feces, urine, tarsal glands, metatarsal, preorbital, forehead, and toes. The male or female in the human species perfume themselves to prepare for an erotic date; aroma is used as a source of attraction but also natural body odor is indicative of affinity and health. The way social insects communicate with each other, as their caste systems, includes tapping, stridulating, rubbings, antenna contact, tastings, strong breathing, and chemical samples. (Wilson 1980, 192)

Zoo-Ludens

Playing is often linked to aesthetics, particularly by Kant's concept of "free play of imagination and understanding" repeatedly proposed in the *Critique of Judgement* to describe aesthetic experience. The relation these two activities have was hightlighted by Schiller in *Aesthetic Education of Man* and more recently by Huizinga in *Homo Ludens*. Koestler (1976) pointed at three kinds of human experience: ludic, aesthetic, and eurekic, mistakenly implying in this trichotomy that playing is not cognitive and that discovery and learning do not involve the aesthetics, nor aesthetics include play and inventiveness, or that these kinds of experiences occur only in humans. Darwin, on the contrary, was well aware of their interrelation:

> How often do we see birds which fly easily, gliding and sailing through the air obviously for pleasure? The cat plays with the captured mouse, and the cormorant with the captured fish. The weaver-bird (*Ploceus*), when confined in a cage, amuses itself by neatly weaving blades of grass between the wires of its

> cage. . . . Hence it is not at all surprising that male birds should continue singing for their own amusement after the season for courtship is over. (Darwin 1882, 370)

Playing is an activity that has been detected among vertebrates like birds and mammals, but there is plenty of evidence that many other species also play in childhood and in adulthood.[25] It has been debated whether reptiles are able to play and Burghardt (1998) argues that at least turtles do have the ability to play. For this reason, mimicry in sticklebacks or transvestite fish that pose as another species or as a female could well be a result of training through playing. Playing is directly related to the learning ability, as in kittens whose movements are always associated with hunting: chasing a string or yarn is analogous to a mouse's tail, a flying object to a bird, and lateral movement of its leg to catch a fish.

At the Fleishhacker zoo, Gregory Bateson discovered in 1952 that monkeys not only fight but play to fight, and they mark it with a special gesture that means "this is play." (Bateson 2000, 533) Chimpanzees, papios, and old world monkeys have an open mouth expression that represents "I am joking," a "playing face" or playful gesture.[26] Marc Bekoff and John Byers mention three types of games among animal species: locomotion, objects, and social. (Bekoff and Byers 1998) Birds do pirouettes, mammals simulate battles, cats and dogs play with balls. The general interpretation of this phenomenon is that it is training for adult life, but such games can also be for pleasure.

From animal species to humans, playing has particularly been a means for training in the struggle for existence. Using Caillois's (1994) categories, we can say that *agon* or competition for sexual partners, food, and territory is characteristic of all sexual species. Games of *ilinx* or vertigo are exemplified by birds diving in flight at great speed to display skill, and the game of "what if" or exploration I termed as *peripatos* is common and natural among all puppies and chicks in their growth.[27] The game of *mimicry* or "as if" and simulation is played by dolphins mimicking the position of seals during sleep or imitating divers cleaning algae from their instruments. (Wilson 1980, 228–29) Mimicry is among the most impressive phenomena in the animal kingdom. Insects camouflage themselves as leaves (*cyclopetra* or *chitoniscus feedjeanus*), non-toxic butterflies mimic toxic ones, fish mimic sand, and the incredible case of wunderpus or mimic octopus who masquerades as other species. (Trivers 1985, figure 16–3, 398) We will see these cases in more detail below.

The Three Thresholds of Perception

Visual assessment in various species may be relatively fine or coarse, as demonstrated by the curious case of moths in the south of France that fly to flowers painted in wallpaper, or as mentioned by Darwin (1871, 317): "the common white butterfly often flies down to a bit of paper on the ground, no doubt mistaking it for one of its own species."

> We have seen that their colours and elegant patterns are arranged and exhibited as if for display. Hence I am led to believe that the females prefer or are most excited by the more brilliant males; for on any other supposition the males would, as far as we can see, be ornamented to no purpose. We know that ants and certain Lamellicorn beetles are capable of feeling an attachment for each other, and that ants recognize their fellows after an interval of several months. Hence there is no abstract improbability in the Lepidoptera, which probably stand nearly or quite as high in the scale as these insects, having sufficient mental capacity to admire bright colours. They certainly discover flowers by colour. The Humming-bird Sphinx may often be seen to swoop down from a distance on a bunch of flowers in the midst of green foliage; and I have been assured by two persons abroad, that these moths repeatedly visit flowers painted on the walls of a room, and vainly endeavor to insert their proboscis into them. (Darwin 1882, 316–17)

Yet if certain beetles and ants are able to recognize each other after months of separation, it is not at all improbable that some species capable of playing and identifying are also capable of admiring bright colors, complex patterns, rhythms, or symmetric compositions, while others barely distinguish the most obvious shapes. (Isingrini 1985; Morel, Meer, and Lavine 1988) Recognizing individuals is possible among animals by what Sebeok defined as the singular proper name (SPN), which labels each individual in its own social environment through various channels: scent, appearance, and sound, even via electrical cues. Moreover, Sebeok proposes the hypothesis that "vertebrates that are capable of individual recognition tend to play as well and vice versa." (Sebeok 1990, 85) We could add that, in addition to the name, they also recognize the surname, e.g. *Lasioglossym zephyrum* bees that detect by smell and only accept their relatives in the honeycomb. (Trivers 1985, 129)

As animals perceive their environment, they also communicate with each other, sending different types of messages: threats; sexual enticements; signs of submission and leadership, alarm and territoriality, food location; invitations to play; appeasement signals; and so on. Since perception begins at basic levels and achieves very complex ones, it is necessary to distinguish three thresholds in perception/interpretation or aesthesis/semiosis, applying heterodoxically Peirce's triadic model of sign processes.[28]

1. Firstness, by *sensation* of environmental information. Cnidarians (corals, anemones, and jellyfish) detect their environment through a variety of sensory organs: cilia composed of nerve cells and touch senses; photoreceptors; statocysts for gravity and balance; and rhopalia, from basic chemical sensors to wide-ranging, sophisticated neuronal receptivity.
2. The second threshold is specific to sexual selection, as it operates through *discernment*, comparing in secondness one individual fitness-signaling to another. Here, perception expands one's body to a second body. In the simplest aquatic animals, where gametes are only released in water, no discernment is necessary. Some species have very low sensitivity and hardly any power of selection, as with male turkeys who are propelled to copulate at the mere sight of a female's head, without a body, even if the head is made out of wood and lacking eyes or beak, while others, such as the peahen, have a very refined or capricious discrimination. (Wright 1994, 47; Trivers 1985)

Kalevi Kull proposes that the mechanism of recognition holds the key to explain what for Darwin was a mystery in the evolution of species, namely, their discontinuity. (Kull 1991, 226–29) Since for Kull the specific partner-recognition system is responsible for the stable range of variability in population characteristics, we could say that such a system stands on this second threshold. Discernment of sexual selection also applies to the recognition by the mother of her offspring through smell, color, or design, as the red spots on the head of gull chicks or by registers pointed out by Sebeok's SPN. (Trivers 1985, 8, fig. I–11)

3. The third threshold can be characterized as social regulatory *regard*, demonstrated when the organism interacts in a group according to its role, obeys the rules of the group, and has a collective mode of sensing. Examples of this third threshold include bird-flock alignments; *Vibrio fischeri* bacteria-quorum sensing; avian and insect swarm behavior; and the grouping of slime mold in pistils for spore migration. Submitting to the group leader or exhibiting dominance in alpha males establishes a social topography that assigns specific roles to each member and maintains a particular collective sensibility.

This threshold is illustrated in the experiment by Sievert Rohwer who, given that dark colors are an index of higher status, colored sparrows' feathers lighter or darker and observed their behavior. The bleached birds became more aggressive because their place had not been acknowledged, while those who were darkened were initially able to deceive others on their status; but when the trick was revealed, these birds were mercilessly attacked for violating the group law.[29]

In short, the first threshold focuses on individual survival, the second on mating and genetic survival, and the third on collective survival. I wonder if these three thresholds could be somehow related to Piaget's three ontogenetic classes of play: sensory-motor in sensation by firstness; symbol or fantasy at the second by discernment between reality and imagination; and games with rules or reflective regard in the third. (Piaget 1999, 113; Sebeok 1990, 89–90)

Three Types of Poiesis: Auto-Poiesis, Phylo-Poiesis, and Onto-Poiesis

I will apply here the same terminological distinction denoting process-study as in semiosis-semiotics: aesthesis-aesthetics, and poiesis-poetics. The first level of this process corresponds to *auto-poiesis* or self-organization in all living systems. The concept of autopoiesis as a condition of life was broadly explained in Francisco Varela and Humberto Maturana's biological approach to knowledge and language (Maturana and Varela 1992) Jakob von Uexküll's *funktionskreis* model points to the continuous cycle of an organism's interaction with the environment, involving perception that propels action, each adjusting the other in an effector-perceptor dynamic. (von Uexküll 1982)

In natural forms, as the case of mollusk's shells or plant leaves, no deliberate interaction from an actor toward a re-actor takes place, but only *auto-poiesis* as characteristic of the self-organization of all living systems. Mollusks' shapes are not produced to be perceived since these are forms that evolve through *auto-poiesis*, according to the constraints of physical laws such as the resistance of matter and energy fields.[30] Their rhythmic patterns are a result of cadence in growing and molecular assemblage and not a re-actor-oriented elaboration. (Portmann 1967; 186)

The second kind of *poiesis* implies an indirect transformation of the species' phenotype across generations due to the females' choice of particular traits through sexual selection: the emphasis here is on form by preference, not spontaneous development. The process of configuration of future generations by sexual selection I will denote as *phylo-poiesis*. This is what Darwin was referring to in the *Origin of Species* when he wrote that "if man can in a short time give elegant carriage and beauty to his bantams, according to his standard of beauty, I can see no good reason to doubt that female birds, by selecting, during thousands of generations, the most melodious or beautiful males, according to their standard of beauty, might produce a marked effect." (Darwin 1859, 89)

In the domestication and breeding of pedigree animals, Darwin finds the clue to explain selection processes in nature that will later lead him to develop the argument of sexual selection in the two impressive volumes of *De-*

scent of Man, the first and insuperable treatise on zoo-aesthetics. This trans-generational modeling of the species by selection is *phylo-poiesis*.

> Just as man can give beauty, according to his standard of taste, to his male poultry, or more strictly can modify the beauty originally acquired by the parent species, can give to the Sebright bantam a new and elegant plumage, an erect and peculiar carriage—so it appears that female birds in a state of nature, have by a long selection of the more attractive males, added to their beauty or other attractive qualities. No doubt this implies powers of discrimination and taste on the part of the female which will at first appear extremely improbable; but by the facts to be adduced hereafter, I hope to be able to shew that the females actually have these powers. (Darwin 1882, 211)

There is a third kind of poiesis, akin to Jakobson's (1960) concept of the poetic function, which emphasizes form over other aspects of a message in a communicative process to provoke certain effects in the receiver. Since this type of poesis involves deliberate elaboration and skills attained throughout a lifetime, it can be termed *onto-poiesis* when shaping or elaborating the form of the message to perfectly fit its receptors. Beyond the spontaneous activity of autopoiesis in the everyday practice of survival, and the species phylo-poiesis by sexual selection, the individual deliberately prepares artifacts and displays by ontopoiesis altering aspects to fit the destinatary.

Onto-poiesis corresponds to a quantum leap of perception when the body opens up to others weaving their *umwelten* together. In this third threshold the effector becomes an actor who directs its action to another perceptor as re-actor (implying what Dawkins and Krebs called "mind reading"). (Krebs and Dawkins 1984) In other words, perception and action processes are no longer just directed to food and security but also involve social survival in seeking territory, information, influence, charm, status, power. The first object-oriented *umwelt* evolves, incorporating also a subject-oriented *umwelt* into overlapping *umwelten* and *innenwelten*, as if the subject would double itself reading or foreshadowing the mind of the other. Messages are constructed to provoke re-actors by adjusting the form of the message to its receptors.

It is here where the zoo-poetic dimension emerges, as soon as form is deliberately modified for the sake of its effect in being perceived by others.

We can now point out that *auto-poiesis* may correspond to the first threshold of perception in sensation as it involves sheer survival, *phylo-poiesis* to the second by discernment in sexual selection, and *onto-poiesis* to the third, by learning a group's conventions as in bird dialects and performance according to rules.

Just as "poetics" refers to the enunciation of the effector or actor, "aesthetics" refers to the perceiver–re-actor (although the first re-actor is the actor itself). In other words, the aesthetic–poetic relationship is characterized by processes of the perceptor–effector and re-actor–actor highlighting signifiers over signifieds and syntax over semantics.

In Figure 3.1, a table that summarizes concepts developed here, we can find correlations that suggest further research possibilities.

Phenomenon	Firstness	Secondness	Thirdness
Threshold	Sensation	Discernment	Regard
Survival	Individual	Generational	Group
Dynamic	Perceptor/ Effector	Actor /Reactor	Tuning or *Abstimmung*
Poiesis	Autopoiesis	Phylopoiesis	Ontopoiesis
Play	Sensorio–Motor Ilinx	Fiction Mimicry	Rules Agon

Figure 3.1.

Zoo-Poetics: Animal Creativity

Aristotle's definition of "poetics" as the study of drama is derived from the Greek word *poiesis* (making or creating) and generally used to refer to the elaboration of formal qualities in artistic works. I therefore propose the term "zoo-poetics" to denote the study of animals' elaboration of particular objects or displays through the enhancement of formal aspects to produce perceptual effect on the re-actor or receiver (as in Jakobson's [1960] sense of the "poetic function"). The difference between a bird's call (semantic) and a bird's song (syntactic), between an animal's doing (praxis) and an animal's displaying (poiesis), between a nest for laying eggs (praxis) and an elaborate bower for impressing the female (poiesis), all of these converge upon and justify the term of zoo-poetics. As proposed before, poiesis branches into three parts: 1) spontaneous organization by self-production in auto-poiesis, 2) species modeling or generational poiesis by sexual selection in phylopoiesis, and 3) individual presentation or display in onto- poiesis.

In line with Hannah Arendt's (1998) distinction between labor and work, we distinguish between labor or praxis and elaboration or poiesis. Labor or praxis is the daily sustenance of the individual that characterizes the first level of autopoiesis as activity for survival. The second level or phylopoiesis takes part in mate selection, reproduction, and care of offspring. Ontopoiesis is the deliberate development and deployment of artifacts or displays that are directed to impact a re-actor either for seduction, distraction of predator, for elevating the emitter's status or securing its protection by the leader and, finally as power display. There is an additional element of fiction or *fictio,* the Latin verb *fingere* meaning shaping, molding, which is the act of feigning, pretending, or presenting something imaginary as if it were real. Animal arts of deception are remarkable poetic displays. Trivers writes that deception and self deception are essential survival mechanisms, not only those for a clear image of the world:

> …[I]f (as Dawkins argues) deceit is fundamental in animal communication, then there must be strong selection to spot deception and this ought, in turn, to select for a degree of self-deception, rendering some facts and motives unconscious so as not to betray—by the subtle signs of self-knowledge—the deception being practiced. Thus, the conventional view that natural selection favors nervous systems which produce ever more accurate images of the world, must be a very naïve view of mental evolution.[31]

Phylo-Poetics

The peacock's amazing tail is not only an obstacle to its own survival but also to Darwin's theory of natural selection, forcing him to write an additional two-volume book on sexual selection. When he asked Mr. Bartlett, who had long experience with birds at the Botanic Garden, if the male tragopan pheasant is a polygamous bird, he received the same answer a daughter would get from her mother when asking about a romantic prospect with a Don Juan: "I do not know, but should think so from his splendid colours." (Darwin 1882, 220) That males with the most colorful plumage and ostentatious displays are polygamous can be explained by the cruel fact that they are all victims, not so much of their own vanity, but of females' preference.

Ronald Fisher attempted to explain this anomaly by what cannot be described otherwise than as a whim or fashion in females' taste. These features are generated by a mechanism he defined as the "runaway process" of preferences where the next generation not only bequeaths the preferred trait to its male offspring but also the preference for that trait to its females increasing in an exponential progression.

Moreover, since these promiscuous peahen-izer males do not invest but a few ejaculating seconds in their offspring, and do not participate in rearing, they can be easily dispensed with or substituted by others. This is the case also of the attractive male silkworm, which can fertilize three females. More alluring aesthetic strategies are generally displayed towards the gender that in the long term invests more energy in the offspring, as Trivers (1971) so acutely argued. Speaking about investments, zoo-poetics thus appears to obey some sort of aesthetic economy where the aesthetic exchange-value of the male's appearance is traded for the metabolic use-value of the peahen's maternal effort in parental investment in Marxists' twofold sense of value. This economy of exchange was confirmed by the observation that, when males incubate the eggs, as in the phalaropes, the roles are reversed, and it is females who exhibit striking colors and are forced to compete. Female marine iguanas are bright and colorful, as they have to compete for the best places on the Galapagos to lay their eggs, which are scarce on the islands. Thus the ability to display and captivate operates not only at the time of sexual selection, but also to ensure a future for their offspring. It seems that the aesthetic, far from being disinterested and useless, has an exchange value as well as a use value, since beauty can be exchanged for maternal energy and labor. In other words, such lavishness is necessary, and such excess indispensable for the reproduction of the species. These assets also participate in a sort of system of substitution. Darwin notices that the ability to sing and the ability to display colors often substitute for each other.

> It is also remarkable that birds which sing well are rarely decorated with brilliant colours or other ornaments. Of our British birds, excepting the bullfinch and goldfinch, the best songsters are plain-coloured. The king-fisher, bee-eater, roller, hoopoe, woodpeckers, &c., utter harsh cries; and the brilliant birds of the tropics are hardly ever songsters. . . . Hence bright colours and the power of song seem to replace each other. We can perceive that if the plumage did not vary in brightness, or if bright colours were dangerous to the species, other means would be employed to charm the females; and melody of voice offers one such means. (Darwin 1882, 371)

In this aesthetic market, the male paradise bird unfairly has, as we shall see, too much aesthetic capital. It has amazing plumage, acrobatic skills to hang upside down from a branch, and an incredible ability to dance and make sounds, all to charm the female. M. J. Ryan distinguishes between indirect selection where a phenotypic trait is chosen for its correlation to the genotype, and direct selection, as in bright colors or sounds that are more easily discernible. (Ryan 1998) He contends that while good gene signaling must be reliable, it does not predict what phenotypic form the new genes should take. He argues that males exploit those perception biases in female preferences, developed by sexual selection, that do not always indicate greater fitness, and thus females can alter the evolution of the traits they find attractive. However, Paul Sherman and Hudson Reeve (1999) question this idea of sensory exploitation because a female thus exploited could eventually be punished through evolution undergoing a reduced reproductive success, and probably become extinct. If sensory exploitation really takes place through the peacock's or pheasant's tail, it seems to me that is not the male exploiting the female in order to reproduce. Rather the opposite seems to be the case, namely the female exploits the male by forcing him to be attractive for her own delectation (although she often appears totally indifferent to his charms), thereby putting him at risk of predators by his conspicuousness. She does so by reproducing these handicaps for her male offspring's sex appeal and for her female offspring's delight.

Onto-Poetics

If the idea of animal aesthetic preferences looked quite odd, as Wallace insisted, then alluding to animal poetics seems totally preposterous. The humorous conceptual work by Vitali Komar and Aleksander Melamid (1999) on "elephant art" exhibited at the Modern Contemporary Art museum in Sydney in 1991, and auctioned for $75,000 at Christie, satirizes the absurdity of this proposition. This was a curatorial reframing of induced animal conduct into a pre-established artistic convention totally alien to an elephant's natural activities. Indisputably what was sold were genuine elephants' paintings (as they involve real paint and real elephants), but the category of "art" is an altogether different matter, a forced anthropomorphic framing. The only art here is Komar and Melamid's conceptual work of going through the whole process to the point of convincing the art world and getting it auctioned at such prices. This is not, however, what we mean by "animal poetics."

The third poiesic process or ontopoiesis, as it involves an individual's elaboration of artifacts or behaviors, can be expressed by visual, acoustic, somatic and (in the human animal) verbal means. The main issue for onto-poetics is not only related to the perception of cues or signals, or the expression of actors, but also to the construction of impressions on re-actors by the deliberate choice of attractive signifiers that communicate factual or illusory realities. Onto-poetics involves passing from the first level of the functional cycle to the second, and the qualitative leap from effector to actor and from receptor to re-actor. Apart from the illusions created by animal mimicry, we can speak of a deliberate construction of imaginary situations by certain species, as in animal cheating, mimicking, and playing.

We may recall that Eco defined semiotics as "in principle, the discipline studying everything which can be used in order to lie." (Eco 1976, 31) Martinelli (2004) addresses precisely this faculty of animal lying in his zoosemiotic approach to aesthetics. He applies Peircean categories to explore the relation between lying, playing, and aesthetics, and classifies animal lying as firstness, playing as secondness, and aesthetics as thirdness. Although playing and lying are indeed very close to art and aesthetics, the linearity of Martinelli's trichotomy is not only anthropocentric but quite arbitrary in constricting aesthetics to conventional arts, placing it at the top of the scheme and hierarchizing these three closely interrelated and multilayered activities.

Beginning his paper on the *Prefigurements of Art*, Sebeok (1979, 4, 8) emphatically rejected the idea that verbal semiosis is a superimposition, so to say, upon a cruder nonverbal semiosis, while at the same time discussing a "superimposed aesthetic function" (somehow echoing Dewey's (1980) idea of "an experience" over everyday experience). I claim, on the contrary, that aesthetics is not superimposed on the ordinary but emerges with it from the very beginning, together with semiosis, at the most basic level of the perceptor-effector functional cycle, and develops differentially through the evolution of each species. What we commonly call "aesthetic" is a matter of degree, not of type or kind, since all species exhibit their own forms of aesthesis depending on the particular condition of their neural system. Sebeok insists on a qualitative difference between language and nonverbal forms of communication (i.e., language developed in specialized brain areas such as Wernicke's and Broca's). The same can be said between art and poetics, not between aesthetics and perception. As language is a purely human type of semiosis, so art is a purely human type of poiesis. This does not cancel the fact that there are semiosic practices in other species, as there are aesthesic ones. Aesthetic delight is a qualified aesthesis, not a superior or superimposed delectation on top of common pleasure. This conception differs from Welsch's (2004) claim for the superiority of aesthetic pleasure over other forms of pleasure as well as from Martinelli's (2004) layering of aesthetics over lying and playing.

Consistent with the idea that art is an institutional phenomenon, as Danto (1964) and Dickie (1974) have argued, one can still propose a zoo-poetics but not a zoo-artistics, as long as one differentiates "art" from "poiesis," the latter an activity from which the former evolved.[32] While art is institutional, conventional, and framed, poiesis is the elaboration of artifacts, displays, or messages with particular emphasis on formal qualities, such as a song to a bird call, or a bower to a nest. Many species make artifacts: bird and fish nests, beaver dams, beehives, spider webs, etc. In all of them, formal aspects are involved, since all construction and communication depends on morphological constraints but not all are dependent on an institution's conventions. When an additional effort is invested to enhance the form by emphasizing color, symmetry, rhythm, or proportion, we are dealing not only with praxis but with onto-poiesis. For Sebeok these are prefigurements of art which he classified as pictorial, musical, kinesthetic, and architectural. (Sebeok 1979, 11) I slightly depart from his classification for reasons I hope are justifiable in proposing the following: visual (pictorial and architectural), musical, theatrical (fiction), and dance (kinesthetic). Homo sapiens sapiens and Neanderthals both practiced onto-poiesis in creating ochre painting, carving bifacial axes, singing, performing, and dancing. (Mithen 2005) This does not mean they were prehistoric artists; what it does mean is that onto-poiesis has a very long evolutionary history that transcends our species. (Bahn and Petit 2013)

In sum, there are two modes of zoo-poetics: One is indirect, *phylo-genetic poetics*, referring, as we shall see, to the visual conformation of the species through many generations as a result of female sexual choice of particular male traits in color or composition. The other is direct, *onto-genetic poetics*, generally performed by males who deliberately construct attractive artifacts like bowers, or decorate nests, for their immediate alluring visual effect. *Phylo-poetics* centers on the genotype, whereas in *onto-poetics*, the phenotype is at stake; one relates to what the individual is, and the second to what the individual does.

Visual Poetics

> The Australian satin bower-bird is the most remarkable of that class which exhibit taste for beauty or for glittering objects out of themselves, that is, beauty not directly personal; collecting, in fact, little museums of shells, gaudy feathers, shining glass, or bits of coloured cloth or pottery.
>
> (Shaw 1866)

Darwin expressed his enthusiasm for the visual charm among animals with multiple and very eloquent phrases: "how wonderfully splendid," "so fabulous," "decorated with beautiful metallic tints," "incredibly colored," "in strongly contrasting colors," "no language sufficient to describe the splendor of the males of some tropical species." (Darwin 1882, 316–17, 400) There is a great variety of horns among insects that have no practical use, and these seem to be devices for aesthetic display by *phylo-poetics*. Darwin mentions the male beetle *Chalcosoma*, which, if magnified to the size of a horse or dog, could be one of the most amazing animals, with its polished black surface and horns. He also mentions the *Elaphomyia* insect (or *Phytalmia* in current biological taxonomy) whose horns probably serve as ornaments because they are pink with black rims and which also have a central line with enormous horns that, like peacock tails, obstruct their normal activities. (Darwin 1882, 280, 297)

In *onto-poetics*, there are numerous examples of animal visual attentiveness to, and elaboration by means of, color intensity, form, composition, texture, brightness, balance, and so on. The various techniques for nest building (hanging, cup-like, buried, basket-like, hollowed out, hidden in a hole, and recycled from other species' nests) among avian species such as weaverbirds, *Malimbus Cassini*, hummingbirds, and the great reed warble, all obey patterns of great elegance, rhythm, and symmetry; the same can be observed in beehives and ant and termite colonies. Even simpler organisms such as the amoeba *Difflugia corona* build incredibly complex nests with a crown at the entrance, which by cell division she passes on to her daughter. (Hansell 2000, 15)

Sara Östlund-Nilsson and Mikael Holmlund observed that three-spined sticklebacks decorate their nests with algae of rare colors or with shiny colorful foil sticks and spangles if available, handicapping the nest by greater visibility but exhibiting it as a signal of potential parental care. (Ostlund-Nilsson & Holmlund, 2003) The Black Wheatear *Oenanthe leucura* even collects a kilo or more of stones in the nest to impress the female.

Yet no species has been found that is more excessive than the Australian bowerbird in nest decoration. These birds build their love nests and decorate them with rare materials, such as colored parrot feathers, bones, bottle caps, leaves, shells, and anything else showy, particularly things that are blue, which they weave among the branches. They even steal materials from their neighbors' nests. Spotted bowerbirds elaborate high decorative grass in the form of arcs touching on the tips, and put pebbles to keep everything in place, marking paths that lead to the nest. These nests or bowers are exclusive for mating since they are placed on the ground, while the breeding nests are in a tree. Both sexes assist in building this love nest, though the male invests much more effort in its elaboration.[33] They place berries of various colors, especially blue, red, and black, and for a fresh addition paint it with berry juice and pink buds of such exquisite taste that John Gould exclaims, "these passages are the most wonderful instances of bird architecture discovered to date."[34]

> The best evidence, however, of a taste for the beautiful is afforded by the three genera of Australian bower-birds already mentioned. Their bowers . . ., where the sexes congregate and play strange antics, are variously constructed, but what most concerns us is, that they are decorated by the several species in a different manner. The Satin bower-bird collects gaily-coloured articles, such as the blue tail-feathers of parrakeets, bleached bones and shells, which it sticks between the twigs, or arranges at the entrance. Mr. Gould found in one bower a neatly-worked stone tomahawk and a slip of blue cotton, evidently procured from a native encampment. These objects are continually re-arranged, and carried about by the birds whilst at play. The bower of the Spotted bower-bird "is beautifully lined with tall grasses, so disposed that the heads nearly meet, and the decorations are very profuse." Round stones are used to keep the grass-stems in their proper places, and to make divergent paths leading to the bower. The stones and shells are often brought from a great distance. The Regent bird, as described by Mr. Ramsay, ornaments its short bower with bleached land-shells belonging to five or six species, and with "berries of various colours, blue, red, and black, which give it when fresh, a very pretty appearance. (Darwin 1882, 413)

Sebeok describes the bower by the orange-crested gardener in the rain forest of New Guinea in absolutely impressive terms:

> The two openings in front of the hut are connected inside by a semicircular passage. The bird has covered column between the two openings with dark moss. It is decorated on one side with blue iridescent beetles, in the middle

> with yellow flowers, and on the other side with broken shells. In front of the bower is a fence plaited from twigs and decorated with brightly colored fruits (sometimes with flowers as well), which forms the boundary of the 'garden.' (Sebeok 1979, 4 n. fig. 2)

In more conventional artistic forms brought to animals, such as painting performed by chimpanzees, Schiller's Alpha or Morris's Congo, for example, the fact that one can find a certain rhythm in colors and strokes may owe more to the natural rhythm of the body movements than to any expressive or communicative poetics on the canvas or paper. As with the "mathematical" horse Clever Hans, who appeared to perform arithmetic operations, the artistry was in the eye of the trainers, who took away the canvases at a certain point when it was pleasing to their sensibility, not the chimpanzees' as Alpha would continue to add paint if allowed, in contrast to Congo.[35] So as with the elephant paintings, not everything colorful or decorated is onto-poetics: it has to arise from the individual's initiative and creativity, according to the species and the individual's own mode of expression.

Musical Poetics

> From ancient references in Eastern literature to the serpent charmers down to modern times, the facts showing that certain animals are gratified by music have been accumulating. He would be a bold man who should say that the birds have no delight in their own songs.
>
> (Shaw 1866)

On the musical powers of various species to seduce the opposite sex, Darwin writes profusely. He stresses that the vocal organs were used and refined for the propagation of the species, such as male insects' stridulating organs, the music of which we find so pleasant when the same note is repeated rhythmically. He adds that male fish produce sounds in the breeding season, as well as frogs and toads among amphibians. The male tortoise makes a sound only in the season of love, and male alligators also roar at this time. Among mammals, usually it is the males who use their voices for the "love call" during this season, and in some cases they remain silent the rest of the year.

We must here distinguish between semiotic and aesthetic acoustics, namely between call sounds and song sounds: calls are generally a brief, effective transfer of information, whereas song, being syntactic and aesthetically dominant, obeys more complex rhythmic patterns and can last a long time. As in human sounds, we must also distinguish between song and instrumental music. Singing is sound that is emitted by passing air through the vocal cavity, while instrumental music uses exterior parts of the body or other objects like sticks or branches to produce sound. Stridulation turns the body into a string instrument as it works by friction; singing converts it into a wind instrument; and shaking membranes by cicadas or the chest beating by chimpanzees turn it into a percussion instrument.

Crickets are said to sing but in fact they stridulate since they do not pass air through the vocal cavity but instead brush one wing against another's combed vein to produce the sound; much like a chord instrument. There are several types of cricket stridulation: the call to the female, which has a high volume to cover large distances and repel other males; the courtship tune when the male is already near the female, which is softer; the aggressive tune, which is used when one male detects another; and finally the copulatory interlude. The females of the *Homoptera* and *Orthoptera* appreciate the musical tones of the male, which have been developed by sexual selection, especially the crickets and grasshoppers from the family *Achetidae*. All of this is described by Darwin (1882).

Male *Magicicada septemdecim* and *Cicada pruinosa* insects produce a drumming sound by vibrating their tymbals, which attracts females. Males also organize singing contests where each contestant takes a turn in a strict order.[36] Manakin birds make various types of sounds by clapping and snapping their wings and body parts:

> [I]t is verified that three of four competing hypotheses of the kinematic mechanisms used for producing *snaps*, namely: (1) above-the-back wing-against-wing claps, (2) wing-against-body claps and (3) wing-into-air flicks, are employed between these two clades, and a fourth mechanism, (4) wing-against-tail feather interactions, is discovered. The kinematic mechanisms used to produce *snaps* are invariable within each identified sonation, despite the fact that a diversity of kinematic mechanisms are used among sonations. (Bostwick and Prum 2003, 3693)

Secondary sexual characteristics of birds are similar to those of insects, including organs for vocal or instrumental music during fertilization season. Darwin quotes Johann Matthäus Bechstein, who raised birds all his life: "That the female canary always chooses the best singer, and that in a state of nature the female finch selects that male out of a hundred whose notes please her most."[37] From the viewpoint of expression, the variety of musical techniques and styles used by birds is remarkable, from the turtledove or sweet nightingale's song, to the drumming of a woodpecker, to the trumpet of geese, ducks, and swans, which appear to be pleasing to everyone:

> Mr. Weir has told me of the case of a bullfinch which had been taught to pipe a German waltz, and who was so good a performer that he cost ten guineas; when this bird was first introduced into a room where other birds were kept and he began to sing, all the others, consisting of about twenty linnets and canaries, ranged themselves on the nearest side of their cages, and listened with the greatest interest to the new performer. (Darwin 1882, 368–69)

More recent research on musical onto-poetics, which emerged during the sixties, can be found in ornitho-musicology. Sebeok dedicates one section of his *Prefigurement of the Arts* to music. He quotes Montaigne who believes, probably rightly, than humans may have learned singing and flute playing from birds. (Sebeok 1979, 19)

Darwin mentions a Mr. Waterhouse who refers to the agile gibbon *Hylobates agilis*, whose musical voice always makes sound intervals in halftones, and where the highest note is a precise octave from the lowest. He adds that "the quality of the notes is very musical; and I do not doubt that a good violinist would be able to give a correct idea of the gibbon's composition, excepting as regards its loudness." (Darwin 1882, 567) Gibbons sing in duet during the morning to express the harmony and stability of their relation. (Wright 1994, 54) Dolphins own an acoustic language consisting of sixteen to nineteen whistles, 60–70 percent of which are common to the species regardless of the region. (Wilson 1980, 229) This means that the remaining 30–40 percent of whistles are a local invention and creation; namely, they are onto-poetic, since they have to be learned, which implies freedom and a degree of improvisation and appreciation. Edward O. Wilson reports on the humpback whale the following:

> The most elaborate *single* display known in any animal species may be the song of the humpback whale *Megaptera novaengliae*. First recognized by W. E. Schevill and later analyzed in some detail by Payne and McVay (1971) the song lasts for intervals of 7 to more than 30 minutes' duration. The really extraordinary fact established by Payne and McVay is that each whale sings its own particular variation of the song, consisting of a very long series of notes, and it is able to repeat the performance indefinitely. Few human singers can sustain a solo of this length and intricacy. The songs are very loud, generating enough volume to be heard clearly through the bottoms of small boats at close range and by hydrophones over distances of kilometers. The notes are eerie yet beautiful to the human ear. Deep basso groans and almost inaudible high soprano squeaks alternate with repetitive squeals that suddenly rise or fall in pitch. (Wilson 1980, 108–9)

> During the breeding season male humpback whales sing long, complex songs that are thought to be sexual displays. Songs consist of a series of themes, progressing in a predictable order that may repeat for several hours. We used a small observation vessel to find singing humpbacks and conduct focal sampling, recording behaviour before, during and after playback. (Strictly speaking, we have evaluated the additional impact that LFA sounds have on a singing whale that is already being followed.) (Miller et al. 2000, 903)

David Rothenberg has researched bird and whale songs for more than a decade in his own personal attempt to perform interspecies music. He asserts that while bird calls are partly innate, simple, and repetitive, their songs are more creative, and variable in pitch and rhythm, "repeating patterns, themes and variations, impressive virtuosic trills and ornaments, scales and inversions." (Rothenberg 2006, 9) He vehemently argues against Zahavi's handicap principle as applied to bird song, insisting that it is performed for the sake of singing itself.[38] The case of the nightingale is quite a case in point, as he sings during long periods unrelated to the mating season or other discernible reasons. Rothenberg goes along the aesthetic disinterest maxim of traditional aesthetics, namely Kant's principle, that "beauty is the form of the finality in an object, so far as perceived in it apart from the representation of an end." (Kant 1790, SS 17) Singing may be a gratuitous act having no ulterior instrumental motive, such as mating, but this does not necessarily imply that it is disinterested, since a bird derives pleasure or a sense of pride and power from it. Disinterest and non-instrumentality are not synonymous. One sings because one can sing, as one writes because one can write. A great investment of energy for the sake of singing itself rather than for some other purpose may be gratuitous, but it is not disinterested. If we derive pleasure from an activity such as looking at colorful pictures, hearing or singing songs, or playing a banjo, this very same pleasure cancels our claim to disinterest. We do it for delight. Singing may be motivated by a desire for feeling delight, and the feeling of delight has intrinsic value. We do not feel delight for some ulterior motive although, for Pinker, delight is evolution's way of paying creatures back for contributing to it. (Pinker 1997, 525)

We can conclude, using Jakobson's (1973) functions of language in a very broad sense, that bird calls have primarily expressive, referential (or self-referential, as in an individual call or scent signature), conative, and even phatic functions. Bird songs are expressive of the state of the emitter; metasemiotic as an index of dexterity; phatic with respect to the hearer; conative in commanding attention and during mating season, and yet they also exhibit the poetic function as dominant because of the relative formal complexity of the message. The difference between the call and the song is that the latter involves poiesis; it is elaborated and has alluring effects, even if directed to the singer itself only. Calls, on the other hand, are a matter of praxis and auto-poiesis towards the conservation of the individual, whereas songs are onto-poiesis, as free play of imagination and ability.

Darwin writes that: "from the deeply-laid principle of inherited associations, musical tones in this case would be likely to call up vaguely and indefinitely the strong emotions of a long-past age." (Darwin 1882, 572) Indeed, these songs may well be Madeleines triggering a Proustian search of lost time back into our animal past.

Theatrical Poetics

> I have a black bantam cock and hen. . . . I have tried him several times with the mirror, he being handsome and having a very pretty rose-comb. He never once pecked at his shadow there but walked mincingly and slowly before it on his toes or drew up a foot as he does when one speaks coaxingly to him.
>
> —(Shaw 1866)

By theatrical poetics, I understand the constitution of a fictional or an illusory reality by deliberately acting in a situation as if it were real. In human theater the spectator agrees to sign with the actors a pact of "poetic faith" or "suspension of disbelief" (as the well known term by Samuel T. Coleridge in 1817) in order to empathically follow fictitious actions as though they were real. He or she thus has a chance of vicariously experiencing the thrill and getting emotionally involved in the events.

In zoo-theatrics this pact occurs between two individuals of the same or different species who agree to act in an imaginary situation as if they were fighting when in fact they are playing (coincidentally this same word, "play" is used in theater and music). However, there can be acting even in cases when only one part, the actor, knows it is illusory, as in cheating, con artists displays, and happenings where the audience participates unknowingly in the event (as opposed to well planned and agreed artistic performances).[39] There is no pact and yet there is a fictitious situation to which the re-actor responds as if it were real. Eco's idea of lying and Martinelli's claim for a relation between animal playing, lying, and aesthetics here converge with neo-Darwinian studies on cheating.

A game of mimicry or of "as if" requires that the actor mind-read the re-actor in order to deceive him. In fact, all four categories in Callois's (1994) taxonomy of games are present in animals: *ilinx,* or games of vertigo in the *Scolopax gallinago* acrobatics; *agon*, in the singing competition of *Cicada pruinosa* insects or in puppies competing for a play object; *mimicry* or imitation, in various species and perhaps also *alea* or games of chance in fishing. Among the types of play characterized by Burghardt, namely locomotor, objects, and social, only the latter can be considered acting. (Burghardt 1998, 6)

Not all acting is playing even if it is theatrical (since it can involve enormous energy expenditure or can be humiliating to the actor), and not all acting is deception (as when a pact of playing is agreed upon). The bird of paradise *Paradisaea* hanging upside down on a branch, flapping its wings and feathers to impress the female, illustrates the effort of exhibiting energetic expenditure as an index of animal quality. These actors are definitely expecting a response from their re-actor or, better, "re-actress," but they also risk failure. Alpha males exhibit animal vitality and strength when leisurely walking among subordinate monkeys to provoke admiration. The leader in a pack of wolves proudly holds up its head, ears, and tail and moves around with slow, secure movements to express status, command subordination, and dramatize his power; by contrast, submissive males hunch, and lower their head and tail. (Wilson 1980, 137) With chimps, some subordinates literally bend over and kiss the dominant's feet. (Wright 1994, 245) This overacting surrender is not necessary for the actual submission but it is dramatized, in this case not to cheat, but to exhibit or publicize this position.

The process of recasting an action and performing it in order to provoke a reaction we may call theatricalization. As in conventional human theater, we can humorously differentiate various genres in zoo-theatrics. We begin with the genre of *burlesque.* Various species like to show off, such as the Abyssinian hornbills *Bucorvus abyssinicus,* which inflate their red neck bladders, wave their wings, and expand their tails in order to proudly make their presence known. When displaying in front of the peahen, peacocks exhibit their tails with great pomposity as if on a stage, displaying themselves for admiration, shaking their feathers and producing a particular noise. The male capercaillie *Tetrao urophasianus* has a naked, yellow esophagus which he inflates to a prodigious size when he courts the female; it becomes almost half of his body size, and he also emits a scratchy, deep, and hollow sound. Darwin describes the capercaillie as follows: "With its neck-feathers erect, his wings lowered, and buzzing on the ground, and his long pointed tail spread out like a fan, he displays a variety of grotesque attitudes." (Darwin 1882, 372)

In the genre of *melodrama*, young chimpanzees overdramatize emotional moods in order to manipulate their mothers for their own benefit. That this is an act rather than an action is confirmed when observing a baby chimpanzee, which may "in the midst of a tantrum glance furtively at its mother or the caretaker as if to discover whether its action was attracting attention."[40] The chimpanzee realizes that his mother cares for him, so he exploits this maternal instinct by exaggerating his expressions, to make it appear as if he were in total distress. I have a video of my son performing exactly the same trick: crying exaggeratedly while he looks furtively to confirm whether he is achieving the desired effect or not. In a scene from *Gone with the Wind*, Scarlett O'Hara also performs this act to manipulate Ashley.

In the *epic* or *heroic* genre, gazelles leap to draw attention from the predator, signaling their quality of excess in energy by wasting it and exposing themselves. This dramatic exhibition dissuades predators from chasing this particular individual, according to Zahavi's concept of the "handicap principle." (Zahavi 1975; Zahavi and Zahavi 1997)[41] Another instance of a dramatized display of energy is the Siamese fighting fish *Betta splendens*, who keep one of their gills upright during a fight to exhibit a breathing handicap. When the male stickleback *Gasterosteus* zigzags to court the female, and wins the fight for her, he increases the brightness of his color, while the defeated male decreases his and literally pales. Shrimps are capable of bluffing and appearing larger than they are to scare away predators, and many mammals when in danger bristle their hair, probably also to appear larger or exaggerate body size. In a territorial dispute, ants raise their body and elevate their head and abdomen to impress their opponent, and stand, as if on stilts, drumming on the rival's body.

We may also find the genre of *comedias de enredos* in the style of Lope de Vega, played by *Lepomis macrochirus* males who simulate being females in order to enter another male's territory and fertilize the visiting female's eggs. According to Trivers, one can find a pseudo-female male trying to steal a naïve male's nuptial gift, and because females recognize and value their cunning, some prefer these pseudo-females to regular males. (Trivers 1985, 407–8)

Another genre is the *thriller*, performed by the *Aspidontus taeniatus* fish, which dances and pretends to be *Labroides dimidiatus*, a fish that maintains a mutualist relation with its host by cleaning parasites from the host's scales. Once the host has been fooled, *Aspidontus taeniatus* attacks and bites it. (Darwin 1882, 326)[42] These performances are not disinterested and have a clear ultimate, sometimes lethal, goal. But just like paintings that were painted to please a patron, or to be sold at an art gallery, they still have a poetic character to attract the re-actor. Instrumentality does not annul poetics as long as it enhances formal aspects to provoke a re-actor.

In the *action and adventure* genre, certain birds, such as larks, plunge downward from the sky at great speeds, not opening their wings until the very last moment in order to dramatize and exhibit their skill. Some can even change the shape of their feathers to produce distinctive sounds while in rapid flight. The *Scolopax gallinago*, *S. wilsonii*, *S. frenata*, and *S. javensis* can fly at high altitudes and then rapidly descend in a zigzag curve with their tail extended, producing a special sound as the air passes over their outer feathers. (Darwin 1882, 376–77) Apart from the playful *ilinx* character of these performances, one can understand them as a pleasurable display of dexterity and excellence. As mockingbirds or canaries with their songs, these birds do acrobatics because they enjoy doing them, simply because they can.

Nominated for best actress in the category of *drama* is the killdeer female bird *Charadrius vociferus*, with her magnificent interpretation of the classic "broken wing act." To distract predators from her offspring she emits a distress call as she flaps her wings, pretending to be wounded. For costume design, the prize goes to the truly Cecil DeMille, glamorous productions of the peacock and the paradise bird mentioned in visual phylopoetics. For best actor, the Oscar goes to the mimic octopus *Thaumoctopus mimicus*, who is capable of impersonating in shape and movement fifteen different species: flatfish, brittle stars, giant crabs, stingrays, jellyfish, sea anemones, mantis shrimps, sea snakes, shells, starfish, and lionfish among others.[43] Cold-blooded species such as amphibians, cephalopods, and crustaceans are able to activate chromatophores, cells that allow color change by reflecting light, as well as by a reshaping of their body according to the species impersonated. For example, when simulating a snake, the mimic octopus buries all but two of its tentacles in the sand and extends them.

Dance Poetics

> In Northern America, large numbers of a grouse, the Tetrao phasianellus, meet every morning during the breeding-season on a selected level spot, and here they run round and round in a circle of about fifteen or twenty feet in diameter, so that the ground is worn quite bare, like a fairy-ring. In these Partridge-dances, as they are called by the hunters, the birds assume the strangest attitudes, and run round, some to the left and some to the right.
> (Darwin 1882, 380)

The indisputable king of pop and international champion of *Dancing with the stars* is Snowball, a male eleonora cockatoo, *Cacatua galerita eleonora*, that is able to dance at any pace with such passion and obvious joy that it has more than eight million views on Youtube, as well as many scientists studying it. Snowball attests beyond any reasonable doubt to birds' musical sense of rhythm and cadence since it perfectly follows the music with grace and enthusiasm, moving each leg at a time and rolling its head from one side to the other, even using his crest to emphasize certain movements.[44] Patel has found experimental evidence that the synchronization of body movements with rhythm can be related to brain circuits associated to vocal capacity. (Patel et al. 2008, 7) If this case were not persuasive enough I will quote the discovery of a crane dance in 1926 by Jack Maclaren, who profusely gathers information on this particular zoo-poetic dance form, and has been beautifully captured by YouTube contributors. Sebeok reports:

> In groups of a score or more they advanced and retreated, lifting high their long legs and standing on their toes, now and then bowing gracefully to one another, now and then one pair encircling with prancing daintiness a group whose heads moved upwards and downwards and sideways in time to the stepping of the pair. At times they formed into one great prancing mass, with their long necks thrust upward; and the wide swaying of their backs was like unto the swaying of the sea. Then, suddenly, as in response to an imperative command, they would sway apart, some of them to rise in low, encircling flight, and some to stand as in little gossiping groups; and presently they would form in pairs or sets of pairs, and the prancing and bowing, and advancing and retreating would begin all over again. . . . [A] group of some twenty mountain chickens of a brilliant orange-yellow color, gathered together in a kind of dance characteristic of these beautiful birds. In the center one of the cocks

> executed the dance-like movements, as he hopped about the open place with wings extended and tail outspread. On the branches of the bushes round about, the others sat and expressed their admiration of the dancer with the strangest sounds. As soon as one cock was exhausted, he joined the spectators, uttering a peculiar cry, and another took his place. (Sebeok 1979, 13–14)

Sebeok mentions that already in 1937 Curt Sachs observed chimpanzees dancing around a pole, or two posts: "As forms, the circle and ellipse around the post, the forward and backward pace; as movements, hopping, rhythmical stamping, whirling, and even ornamentation for the dance." Sachs concludes as follows:

> If the dance inherited from brutish ancestors, lives in all mankind as a necessary motor-rhythmic expression of excess energy and of the joy of living, then it is only of slight importance for anthropologists and social historians. If it is established, however, that an inherited predisposition develops in many ways in the different groups of man and in its force and direction is related to other phenomena of civilization, the history of the dance will then be of great importance for the study of mankind.[45]

I beg to differ since, on the contrary, dance as a rhythmic expression of excess energy and of the joy of living is of utmost importance for each and every human being, especially if we realize how long are the roots into our species' aesthetic past.

The manakin moon-walking dance, the cockatoo's dancing for sheer joy, the bowerbird dancing before the pergola, and the bird of paradise's rhythmic displays are among the numerous cases of dance poetics among animals. The cockatoo dances alone, manakins may dance in duets, and cranes dance in pairs or groups with rhythmical and reiterative patterns.[46]

I have argued that animals are not only capable of *aesthesis* in the perception and appreciation of formal patterns, colors, and rhythms, but of *poeisis* as well, since they are able to create and display objects or performances according to formal criteria with the goal of generating a perceptual, fictional, or attraction effect in the receiver. In contrast to "elephant art," jokingly extrapolated by Komar and Melamid, zoo-poetics must be read solely in terms of their authors' initiatives: An elephant does not have the urge to paint, whereas a bowerbird willingly decorates its pergola with berry juice and other colorful items and contemplates his work, walking forward and backward to better appreciate the effect. While art is an institutional practice, and can consequently only be a human pursuit according to certain conventions, zoo-poetics is an activity found in many species to both please and take pride in by attesting to their ability to perform, charm, or deceive others.

To sum up, recapitulation of some terminological distinctions is convenient in dealing with zoo-aesthetics. Aesthetics, or better bio-aesthetics, as the study of the processes of aesthesis in a wide spectrum of live beings, is linked in origin and operation to those of semiotics. Since aesthesis, together with semiosis, is at the root of life systems, the perceptor-effector cycle of von Uexküll's functional model is pertinent also at the second level of social interactions, which turn the effector into an actor and the perceptor into a reactor. Observing animal communication from these two levels, a differentiation between cues or indicators and signals or displays becomes evident.[47]

Another terminological result of previous work is the distinction of the three thresholds of perception, derived from the Peircean categories of firstness, secondness, and thirdness. When applied to aesthetics, these categories involve individual sensation, mate discernment, and social regard.[48] These in turn are linked to the three forms of poiesis: auto-poiesis, or self creation in individual survival, phylo-poiesis, as indirect modeling of subsequent generations by mate selection, and onto-poiesis, as creative elaboration of artifacts or displays by each individual through various registers to affect others. Here, I wonder, given the ability of the deliberate activation of chromatophores, whether some features, such as feather designs, could have originally been onto-poiesic and later became fixed by natural selection. In this case, not only the display but the arrangement is the creature's direct and indirect design.

I must insist on the importance of approaching aesthetics as a natural outcome of evolution and not as a supernatural product of trained inspiration or exclusively human result of culture and genius. Using spatial metaphors, aesthesis is an *under*lying rather than a *super*imposed process, and it can be heightened or intensified not only through art, as described by Dewey's concept of "an experience," but in everyday life. Aesthesis proper to each species' sensorial world in its own terms, springs forth through an excess, refining our perception of color, composition, variation, symmetry, and rhythm: an excess indispensable to evolution.

One of the main problems in evolutionary aesthetics is the tension between Darwinian instrumental approaches and the traditional, specifically Kantian aesthetic principle of disinterest and the intrinsic value of beauty. Any non-utilitarian display of energy, such as the crane's dance or the lyrebird's all-night song, which lasts even beyond the mating period, requires an explanation. We have seen that acting, as with the killdeer mother's performance, can be a matter of life and death, as the legendary storyteller Scheherazade's stories also were invented as a matter of life and death, not demeriting the narrative's imaginative, creative, and literary qualities. A bower exceeding a nest, a proud theatrical display, a graceful dance, or an inspired skillful song, all thrust an individual beyond itself by onto-poiesis, and a species by phylo-poiesis. In singing, as in dancing, in decorating and in acting, there is an excess that overflows necessity, a dimension of gratuitousness that opens the gates of freedom and attests to our creative role in evolution, as both creatures and creators.

IMAGINING: ANTHROPOAESTHETICS

With embroidered charro outfits and galloping on their horses, Pedro Infante and Jorge Negrete, prototypical beaus of Mexican golden age cinema, sing to woo their females just like birds while displaying their plumage and prancing in courtship. Perfect illustrations of our continuity with the animal kingdom, they seem to have taken note when Darwin said that "The courtship of animals is by no means so simple and short an affair as might be thought. The females are most excited by, or prefer pairing with, the more ornamented males, or those which are the best songsters, or play the best antics." (Darwin 1882, 214)

Not surprisingly, in addition to movies and dramas, soap operas are a detailed catalog of Darwinist topics, Dawkinsian in particular, by bluntly proving his "selfish gene" thesis in the recurrent topoi of the drama. The plot in telenovelas is always about keeping track of genes obsessed with replicating themselves and selecting the fittest. Beau and heroine will have to signal their fitness not only by aesthetics but by perseverance, deploying costly signals as they endure their suffering throughout the telenovela. The main characters of the selfish gene soap will overcome the evilness and jealousy of the "less fit" to consolidate the desired happy ending at the altar which symbolizes the promise of the reproduction and improvement of the species. The story revolves as if on a genetic map around who is who's parent, who will mate with whom, what the emotional and physical cost will be to copulate with the alpha female or male, and when to disclose the genetic origin of the main character. Hence the close-up camera technique flaunts an inventory of desirable phenotypes in the leading actors: their perfect teeth, symmetrical nostrils, delineated eyebrows and long eyelashes, jaws and cheekbones, male developed biceps, elongated, strong, and proportionate limbs, nubile female breast, soft skin, and waist-hip ratio of 0.7 males and 0.9 females.

The ranchero song genre puts on display with more-or-less poetic quality a repertoire of emotions strictly linked to the selfish gene's whims and mating compulsion in all its variants: possible and impossible, probable or frustrated and subsequent reactions of jealousy, resentment, revenge, shame, euphoria, emotional blackmail, self-pity, and guilt.

With these examples, one would be forced to capitulate to the mating mind thesis by Geoffrey Miller (2001) for explaining culture. However, if it is true that among humans there are individuals who are obsessed with mating, other species like some birds are far more subtle as they sing after, not before, mating. This behavior seems to indicate that birds may have a wider and more sophisticated horizon of creativity and confirm Kant's thesis on pure aesthetic disinterestedness. The case of the nightingale that sings all night without addressing any particular female and invests a significant metabolic cost losing weight as a result, illustrates the sheer pleasure of singing for and by itself. (Rothenberg 2006, 41)

There is a concept that has become academically incorrect but that should be nonetheless reconsidered in relation to our species. I am referring to the concept of "human nature" because only by assuming it can we present the case or the claim of an anthropo-aesthetics. Marc Hauser (2009) points at four characteristics of this "humanicity" referring to the oneness of the human species: 1) capacity of generative computing with recursion and combinatorial elements, 2) the promiscuous combination of ideas, 3) mental symbols, and 4) abstract thinking. The simple fact of our morphological homogeneity as human species already implies that there is such thing as human nature. Evolution has shaped us with human legs and arms, human eyes and ears, with glands and ability to cultivate culture. If we could speak of the essence of humanity, it is right here in our body that requires the same type of enzymes to digest, of neurons to think, proteins for growth, which enjoys when endorphins flow through our system and supports our body when walking on the soles of our feet. From infancy we prefer sweet to bitter and nurturing to toxic, except for cultural learning that teaches us to enjoy bitter drinks such as coffee and alcohol or toxic substances such as drugs for their stimulating effects on the body.

On anthropo-aesthetics, Darwin writes of the remarkable fact that:

> In most, but not all parts of the world, the men are more ornamented than the women, and often in a different manner; sometimes, though rarely, the women are hardly at all ornamented. As the women are made by savages to perform the greatest share of the work, and as they are not allowed to eat the best kinds of food, so it accords with the characteristic selfishness of man that they should not be allowed to obtain, or use the finest ornaments. Lastly, it is a remarkable fact, as proved by the foregoing quotations, that the same fashions in modifying the shape of the head, in ornamenting the hair, in painting, tattooing, in perforating the nose, lips, or ears, in removing or filing the teeth, &c., now prevail, and have long prevailed, in the most distant quarters of the world. It is extremely improbable that these practices, followed by so many distinct nations, should be due to tradition from any common source. They indicate the close similarity of the mind of man, to whatever race he may belong, just as do the almost universal habits of dancing, masquerading, and making rude pictures.
>
> Having made these preliminary remarks on the admiration felt by savages for various ornaments, and for deformities most unsightly in our eyes, let us see how far the men are attracted by the appearance of their women, and what are their ideas of beauty. I have heard it maintained that savages are quite indifferent about the beauty of their women, valuing them solely as slaves; it

> may therefore be well to observe that this conclusion does not at all agree with the care which the women take in ornamenting themselves, or with their vanity. (Darwin 1882, 577)

As homo sapiens sapiens we also partly encompass the homo economicus, semioticus, politicus, juridicus, sensitivus, or aestheticus that we inherited from our ancestors and pass on to our posthominida descendants. There is no one in our species who did not take, in one form or another, initiatives that deal with the exchange of goods, of signs, of power, of laws or preferences. We have particular ways at different levels of sensitivity that characterize us as a species, as groups, and as individuals.

According to Aztec mythology, the human body was corn kneaded with blood and mud, and Genesis describes it as clay blown by the wind or spirit of God. Four thousand years later, Pinker describes this body as "a fantastically complex machinery-optical analyzer, motion guidance systems, simulations of the world, databases on people and things, goal schedulers, conflict-resolvers, and many others." "The stuff of life turned out to be not a quivering, glowing, wondrous gel, but a contraption of tiny jigs, springs, hinges, rods, sheets, magnets, zippers, and trapdoors, assembled by data tape and whose information is copied, downloaded and scanned." (Pinker 1997, 19, 22) With his typical eloquence, Pinker's mechanical metaphor of the human body may not seem very friendly, yet watch your fingers typing on the keyboard, how efficiently we bend our knees when climbing up the stairs, how accurately we focus our eyes on the clock when we wake up in the morning; the mechanical precision of these devices is astonishing. We are a machine indeed, yet one that also doubts and sweats and trembles with excitement, a machine designed in the image and likeness of our miraculous evolution.

Darwin states that "My object in this chapter is solely to shew that there is no fundamental difference between man and the higher mammals in their mental faculties." (Darwin 1871, vol 1 35.) Sixty-nine pages later he adds that this difference is one of degree rather than of kind:

> Nevertheless the difference in mind between man and the higher animals, great as it is, is certainly one of degree and not of kind. We have seen that the senses and intuitions, the various emotions and faculties, such as love, memory, attention, curiosity, imitation, reason . . ., of which man boasts, may be found in an incipient, or even sometimes in a well-developed condition, in the lower animals. (Darwin 1871, 104)

It is peculiar, then, that Darwin would find a larger gap in the taste for beauty between an educated man and "lowest savages" than between humans and animals.

> When, however, it is said that the lower animals have a sense of beauty, it must not be supposed that such sense is comparable with that of a cultivated man, with his multiform and complex associated ideas. A more just comparison would be between the taste for the beautiful in animals, and that in the lowest savages, who admire and deck themselves with any brilliant, glittering, or curious object. (Darwin 1882, 211)

Passionate Humans

For Ekman (2003), the pioneer researcher in facial expression after Darwin, there are six universal emotions present in all cultures: happiness, sadness, disgust, fear, anger, and surprise (seventh added in 2003, contempt). A human being uses on average 150 to 200 nonverbal gestures. (Wilson 1980, 94) Expression of emotions is innate, because even children born deaf and blind express emotion in their faces.

Darwin argues that among quadrumana (old description of hominids who had four feet) males had vocal organs more developed than females and could cover an octave, probably to seduce them.

> So little is known about the use of the voice by the Quadrumana during the season of love, that we have no means of judging whether the habit of singing was first acquired by our male or female ancestors. Women are generally thought to possess sweeter voices than men, and as far as this serves as any guide, we may infer that they first acquired musical powers in order to attract the other sex. But if so, this must have occurred long ago, before our ancestors had become sufficiently human to treat and value their women merely as useful slaves. The impassioned orator, bard, or musician, when with his varied tones and cadences he excites the strongest emotions in his hearers, little suspects that he uses the same means by which his half-human ancestors long ago aroused each other's ardent passions, during their courtship and rivalry. (Darwin 1882, 573)

The relationship between emotions and reasoning is an old topic that Plato in *Phaedrus* 246a–b tried to explain through the allegory of the chariot (the soul) led by the charioteer (intellect). The chariot is pulled by two horses, one white with a long thin neck, docile and good strain (morality and reason) and a black short-necked, rebellious, obeying only lashes (the passions, instincts, and impulses).

Yet, on the contrary, emotionality is the invaluable guide for survival, as has repeatedly been proven by studying prefrontal cortex lesions. Damasio (1994, chapter 1) cites the case of Phineas Gage whose ventromedial prefrontal cortex was injured from side to side by a beam during an accident. He describes the consequences of that injury by observing the close link between emotion and decision making in Phineas's inability, and patients with similar lesions, to decide the simplest actions, such as going in or out of the shower.

To Francisco Varela, knowledge is tinged with emotional, relational, postural flow. (Varela, Thompson, and Rosch 1992) Knowledge is intentional or directed, for all perception is injected with emotion. In this line, akin to James's pragmatism, emotion is not mental but a physical state that paints environment perception with a certain tone. An unfortunate mental patient of William James expressed most eloquently, the importance of emotional colors that flow daily life by their absence:

> Every function, every action of my life remains, but deprived of the feeling that belongs to it, of the enjoyment that should follow it. My feet are cold, I warm them, but gain no pleasure from the warmth. I recognise the taste of all I eat, without getting any pleasure from it. . . . My children are growing handsome and healthy, everyone tells me so, I see it myself, but the delight, the inward comfort I ought to feel, I fail to get. Music has lost all charm for me, I used to love it dearly. My daughter plays very well, but for me it is mere noise. That lively interest which a year ago made me hear a delicious concert in the smallest air their fingers played,—that thrill, that general vibration which made me shed such tender tears,—all that exists no more. (James 1884)

James held and developed a body-based theory of emotions. He claims in his classic text *What Is an Emotion?*

> Our natural way of thinking about these standard emotions is that the mental perception of some fact excites the mental affection called the emotion, and that this latter state of mind gives rise to the bodily expression. My thesis on the contrary is that *the bodily changes follow directly the* PERCEPTION *of the exciting fact, and that our feeling of the same changes as they occur* IS *the emotion.* Common sense says, we lose our fortune, are sorry and weep; we meet a bear, are frightened and run; we are insulted by a rival, are angry and strike. The hypothesis here to be defended says that this order of sequence is incorrect, that the one mental state is not immediately induced by the other, that the bodily manifestations must first be interposed between, and that the more rational statement is that we feel sorry because we cry, angry because we strike, afraid because we tremble, and not that we cry, strike, or tremble, because we are sorry, angry, or fearful, as the case may be. Without the bodily states following on the perception, the latter would be purely cognitive in form, pale, colourless, destitute of emotional warmth. We might then see the bear, and judge it best to run, receive the insult and deem it right to strike, but we could not actually *feel* afraid or angry. (James 1884)[49]

Current research in molecular biology empirically confirms James's intuition on the physicality of emotion. We can all recognize the corporeality of anger by the secretion of adrenaline when in conflict with another vehicle driver or when hearing unpleasant news, as the body suddenly feels shaken and propelled to react. Furthermore, the secretion of hormones encourages us to exhibit our hormonal status by the tone and dedication of our apparel.

According to Paul MacLean, the different layers of the human brain correspond to different evolutionary stages from the brainstem, which regulates the automatic functions such as temperature, respiration, and blood-flow, the limbic or emotional system, to the brain cortex in which, among other functions, are the visual cortex, motor, prefrontal, in addition to consciousness, memory, attention, language, thought.[50] In the frontal lobes of the cerebral cortex are numerous opioid receptors connected to the amygdala in the limbic system. Areas of pleasure and pain are related to the emotional life and our reptilian brain busy in eating, fighting, fleeing, and fornicating.

The fact that opioid receptors are densely present in the limbic system of the organism or emotional circuits—approximately 85 percent to 95 percent of the neuropeptide receptors are located in limbic structures in the brain (amygdala, hippocampus, and limbic cortex)— leads Pert to the conclusion that the peptides are emotion molecules, especially enkefalin or endorphin. Pain thresholds and defensive reactions are in the periaqueductal gray area (PAG), as well as the arching of the females from pleasure during orgasm in the nucleus accumbens. Emotional responses related to memory, learning, behavior and affections spring forth from the limbic system. Peptides can be found from the neocortex to the limbic system or emotional brain, but they also are secreted by glands and the immune system.

> [I]t was hard to grasp the idea that the ligand-receptor system represented a second nervous system, one that operated on a much longer time scale, over much greater distances.... Especially difficult to accept was that this chemical-based system was one indisputably more ancient and far more basic to the organism. There were peptides such as endorphins, for instance, being made inside cells long before there were dendrites, axons or even neurons—in fact, before there were brains. (Pert 1999, 26)

This finding involved the transformation of the electric model of the brain to reinforce the idea of the chemical brain by semiosic molecular flows that travel throughout the body to connect to its cellular receptors for survival or for death, in cases when the message is wrong (as viral interference in cancer and autoimmune diseases). According to Pert peptides are precisely the molecular confirmation of the unconscious formulated by Freud. (Pert 1999, 141) In fact for Pert the mind lies in what maintains the operation of the web and unconsciously coordinates organs and cells. I have named this web *significans* (in chapter 2, section 4) referring to the "intelligence as information that controls all systems and creates behavior." So the mind does not rule the body, but rather *is* body, its outward manifestation in physical space. "Emotions are at the nexus between matter and mind, going back and forth between the two and influencing both." According to Pert, they are what links body and mind, namely the bodymind. (Pert 1999, 185–89)

There is a large concentration of neuropeptide receptors at sites where information reaches the nervous system through the five senses—sight, sound, taste, smell, and touch. (Pert 1999, 142.) The memories and emotions are stored not as much in the brain as in this psychosomatic web. Cells of pancreas, spinal cord, spleen, lymph nodes, white cells are in constant communication with various parts of the body. The way lymphocytes tend to certain neuropeptides in chemotaxis exemplifies the reality of aesthesis at a cellular level, but also the ability to manufacture chemical information that can regulate mood or emotions. (Pert 1999, 183)

For Steven Mithen (1999), culture lies less in the elements of the brain than in the mental integration of various intelligences by projecting knowledge of one another. In a flexible version of the concept of the modular mind as a Swiss army knife given by Fodor (1983), Mithen argues that there are four types of intelligence: natural or environmental, technical, social, and linguistic. Carving a stone like in Oldowan artifacts, integrates natural intelligence on environmental objects (stones, trees, rivers, plants) with technical intelligence to modify them. Also, if the stone is an element to generate status or seduce a mate, both intelligences are projected into a third, social intelligence, which in turn can be integrated into the fourth to the extent that such stone is used as a sign.[51]

Disfigured Humans

If David Hume would really have wanted to find out what the standard of taste is, he should have embarked on a journey across five continents to verify this. Humboldt did travel to two, Europe and America, and drew such interesting conclusions that Darwin cites his observations:

> "If painted nations," as Humboldt observes, "had been examined with the same attention as clothed nations, it would have been perceived that the most fertile imagination and the most mutable caprice have created the fashions of painting, as well as those of garments."
>
> In one part of Africa the eyelids are coloured black; in another the nails are coloured yellow or purple. In many places the hair is dyed of various tints. In different countries the teeth are stained black, red, blue, &c., and in the Malay Archipelago it is thought shameful to have white teeth "like those of a dog." (Darwin 1882, 574)

> The senses of man and of the lower animals seem to be so constituted that brilliant colours and certain forms, as well as harmonious and rhythmical sounds, give pleasure and are called beautiful; but why this should be so, we know not. It is certainly not true that there is in the mind of man any universal standard of beauty with respect to the human body. It is, however, possible that certain tastes may in the course of time become inherited, though there is no evidence in favour of this belief; and if so, each race would possess its own

> innate ideal standard of beauty. It has been argued[75] that ugliness consists in an approach to the structure of the lower animals, and no doubt this is partly true with the more civilised nations, in which intellect is highly appreciated; but this explanation will hardly apply to all forms of ugliness. (Darwin 1882, 584)

Hearne, who lived many years with American Indians, writes: "Ask what the northern Indian beauty (as Voltaire's imaginary toad) and he will answer: a wide, flat face, small eyes, high cheekbones, three or four wide black lines through each cheek, a short forehead, a large broad chin, a clumsy hook nose, red skin, and breasts hanging down to the belt."[52]

Westerners were seen as human bodies with beak noses. Therefore, among the inhabitants of Tahiti, to be called "long nose" is considered an insult and they compress the noses and foreheads of their children to embellish according to their criteria. So it is with the Malays of Sumatra, the Hottentots and the natives of Brazil. He cites the case of blacks who admire very dark skin and feel horror of whiteness as demons and spirits are white, and as an index of poor health. (Darwin 1882, 579)

Darwin mentions skull modifications from ancient to modern times in many cultures to exaggerate a natural and admired peculiarity. "Many American Indians are known to admire a head so extremely flattened as to appear to us idiotic." (Darwin 1882, 583–84) He adds that:

> [T]he natives on the north-western coast compress the head into a pointed cone; and it is their constant practice to gather the hair into a knot on the top of the head, for the sake, as Dr. Wilson remarks, "of increasing the apparent elevation of the favourite conoid form." The inhabitants of Arakhan "admire a broad, smooth forehead, and in order to produce it, they fasten a plate of lead on the heads of the new-born children. On the other hand, "a broad, well-rounded occiput is considered a great beauty" by the natives of the Fiji Islands. (Darwin 1882, 583)

The relativity of taste is expressed in particular with regard to hair which in some cases is admired for its abundance while others are shaved down to the eyebrows, as among African groups and certain customs of Paraguay because they say they do not want to look like horses. (Darwin 1882, 580) Instead, hirsute ethnic groups pride themselves on their shaggy hair and beards and exhibit them. He mentions many different customs regarding the teeth:

> The natives of the Upper Nile knock out the four front teeth, saying that they do not wish to resemble brutes. Further south, the Bakotas knock out only the two upper incisors, which, as Livingstone[47] remarks, gives the face a hideous appearance, owing to the prominence of the lower jaw; but these people think the presence of the incisors most unsightly, and on beholding some Europeans, cried out, "Look at the great teeth!" The chief Sebituani tried in vain to alter

> this fashion. In various parts of Africa and in the Malay Archipelago the natives file the incisors into points like those of a saw, or pierce them with holes, into which they insert studs. (Darwin 1882, 575)

Baker comments in 1867 that in Arab countries no beauty can be perfect until the cheeks or temples have been injured, and in South America, "accuse a mother of indifference to her children if she did not use artificial means to deform their calf according to fashion of the place."[53] Each ethnic group has its own standard of beauty. "Arab women of the Upper Nile occupy about three days to fix their hair, never imitating other tribes," but simply compete among them to be the superlative of their own style. "[They] only admire what they're used to, but crave to see each feature a little more developed." (Darwin 1882, 583–84)

> Much has been written on evolutionary aesthetics about the human species, for example Grammer and Thornhill who investigated through computerized faces the symmetry and average shape in features considered most attractive. Nancy Etcoff has studied the social benefits of those who have certain anatomical and facial features. (Etcoff 2000)

Preference surveys have shown that when women are fertile they prefer males with more symmetrical faces than when they are not. (Grammer and Thornhill 1994)[54] The result suggests that females may have evolved two different criteria in their preferences for males, some to breed and others to nurture. In fact, blood tests show that in certain urban areas a quarter of the children are not biological offspring of the registered parent.[55]

Piercing of nose, mouth, lips, ears, and septum are common in varied communities. Thus began a diversification of ornaments that not only has sexual selection purposes but signals skill, tribal distinction, and rank. The topic of sex appeal in animal species has been approached from several hypotheses. One of them is the "multiple fitness model" whereby there are several dimensions in the attractiveness, "the multiple message hypothesis" which states that each ornament signals or indicates a specific property of the individual's fitness, and "redundant signal hypothesis," where the combination of various ornaments gives a better estimate of the status of the wearer. (Moller and Pomiankowski 1993)[56] Although these authors take into account only the animal or the nest's physical appearance, we should also consider the attitude displayed in seduction such as toughness, grace, poise, drive, volume, pace, and, in humans, the selection of vocabulary and linguistic world evoked by the individual, the semantic wealth, and syntactic accuracy.

Anthropo-Poetics and the Evolution of Art

We know that organisms evolve, but does art evolve? Is an artifact capable of evolving? If the term is used in a metaphorical sense, one can take a poetic license and assert it, but in a literal sense it would imply a process of random variation and selective retention of traits in the replication of art objects. What then are the traits that are replicated? For what reasons would certain artistic traits be replicated and not others?

If we take Baumgarten's definition of aesthetics as *scientia cognitionis sensitivae* or theory of sensitive knowledge, an evolutionary approach would require acknowledging that both knowledge and sensitivity are products of evolution. However, the path followed in this discipline for two and a half centuries has been to equate aesthetics with the theory of art, from idealistic and romantic aesthetics to contemporary analytic aesthetics. The newly developed evolutionary aesthetics also takes this synonymy for granted to account for an artificial phenomenon such as art, but in terms of a natural process of evolution. The strategies used to date have been understanding art as: an instinct (Dutton 2009), a mechanism to adapt to the environment (Dissanayake 2007), cognitive engineering (Donald 2006), exaptation or transformation of an adaptation initially with another purpose (Gould 1996), a means for pleasure (Welsch 2004), a reward for collaborating with evolution (Pinker 1994, 1997), or a strategy for sexual selection in humans (Miller 2001). We will review some of their main proposals.

Denis Dutton voluntarily placed himself in the predicament of proving that art is an instinct, being assured that the consistency of Steven Pinker's (1994) hypothesis that language is an instinct would guide him in the implicit syllogism of the sort of: all language is an instinct, all art is a language, therefore all art is an instinct. Dutton's deductive reasoning follows the classic syllogism but the use of the term "language" of the minor premise is not literal but metaphorical, so the conclusion is false. Some speak English, Scottish, or Spanish, but to my knowledge, nobody speaks Artish. Language is a noun to denote the activity of languaging which involves well-defined brain regions in Broca's and Wernicke areas, subcortical regions and the putamen, nucleus caudate, pre-motor areas, inferior frontal gyrus and superior temporal.[57] To date there has not been localized a specialized module or area for art in the human brain.

Referring to art as an instinct has several immediate logical consequences: 1) if there were an instinct for art in the human species we would have to explain why we are not all artists. If it is an adaptation 2) how it is transmitted, 3) which are the proximate and ultimate causes of the evolution of artistic traits, 4) what purpose does it serve, 5) how can it be observed in each individual of the species, 6) where is its organic support, 7) how does it contribute to fitness, 8) what are its antecedents in other species, and so on.

Dutton does not offer a clear definition of such an instinct or how it operates. He decides to choose his examples from nothing less than conceptual art in Duchamp, Manzoni, and Komar and Melamid, artists who owe the meaning of their work fully to the art world's conventions, which is the art legitimating institution that does not even depend on the sensory value or beauty, as argued by Binkley (1977). What, then, that instinct is, and how it works, remains unexplained.

That art is produced ultimately for the purpose of seduction is the argument put forward by Geoffrey Miller (2001). With no significant mediation by culture, Miller sees the mind as an organ that evolved primarily because of the need to seduce and compete with rivals to maximize fitness. Contrary to the Freudian concept of art as sublimation of sexual libido, where energy is the force that the individual channels to achieve expression in artistic sublimation, for Miller art and the evolution of the mental ultimately point to intercourse. If, for Freud, sexuality is at the service of art, for Miller art is at the service of sex.

What really does allure a woman? Sweet and loving words uttered with a captivating voice. All cultures have composed lyric poetry, rhymed and rhythmic expression of love. Lyrics are the aestheticization of sexual desire, evidence of the close link between the artifices for seduction and the evolution of the biblical maxim "be fruitful and multiply." Since then poetry for love has been composed from David's *Psalms* or King Solomon's *Song of Songs* to Greek lyrical poetry by Sappho or Pindar and his Roman followers like Cornelius Gallus, medieval Persian poets and Sufis to the legendary Occitan troubadours who proverbially sing to love. And indeed, today the multibillion-dollar music industry exploits this loving instinct with very different expressive qualities. As Marx said, with the production of the object there is also a production of the subject who will use it, so there is a correspondence of a finer or coarser poetic production, with the more refined or coarse taste in the audience who appreciates it.

Miller's thesis seems to be confirmed by the mass production of art in the service of sexuality and fitness, but the poet is now made to seduce not a lady, but the mass of consumers who will buy his albums. The goal now is seducing the consumer to maximize the artist's and agents' monetary, rather than genetic, capital although such capital can attract many prospects in addition to sexual prospects.

The erotic drive, however, can reach beyond sex, since poetry is inspired not only for the sake of courtship or the release of the poison of resentment and jealousy. The Homeric epic of war describes the heroism of the protagonists, the thrill of battle, the mortal wounds of betrayal, the value of loyalty and greatness among so many other human passions. These feelings are also anchored in the evolution of tribal loyalty, affection for mother or child, and admiration for the leader or desire for status and honor.

Art is the daughter of leisure rather than of lust, as the latter craves to satisfy the body immediately, whereas leisure is in no hurry and can play with the abandonment and fantasy that characterizes art and children. Artistic production is the ostentation of excess of available time, playful energy, and sometimes even of talent. As we shall see, Pinker's hypothesis about art is provocative and very unflattering to the art world because it shows that art is a parasitic byproduct resulting from other cognitive functions that are adaptive. He writes:

> We enjoy the strawberry cheesecake, but not because we evolved a taste for it. We evolved circuits that gave us trickles of enjoyment from the sweet taste of ripe fruit, the creamy mouth feel of fats and oils of nuts and meat and the coolness of fresh water. Cheesecake packs a sensual wallop unlike anything in the natural world, because it is a brew of megadoses of agreeable stimuli which we concocted for the express purpose of pressing our pleasure buttons. Pornography is another pleasure technology. (Pinker 1997, 525)

For Pinker, music is an auditory cheesecake of concentrated patterns of emotions and intervals and condensed movements of tension and release. He adds that art is a third button for pleasure.It is tempting to test Pinker with the trap encountered by traditional theories of art, namely, explaining John Cage's 4'33" or Manzoni's *Merda d'Artista* none of which appear to be a strawberry cheesecake.

Let's review the hypothesis of art as megadose of pleasurable stimuli derived from environment information to which we tune to survive. Pinker's position is related to a cognitive hypothesis which explains that satisfaction is derived from a good fit of perceptual functions to the environment. According to it, we are evolutionarily designed to feel dissatisfaction with scenes of blurry or hazy environments, and attracted by others that express order, color, vibrancy, and well-marked areas. He states that "perhaps the melodies are pleasing to the ear for the same reason that symmetrical, regular, parallel, repetitive doodles are pleasing to the eye. They exaggerate the experience of being in an environment that contains strong, clear analyzable signals from interesting and potent objects." (Pinker 1997, 536)

Turner's paintings of vaporous clouds and tumultuous waves, as well as Tapies, de Kooning, Burri, or Pollock's informalist paintings contradict Pinker´s hypothesis. In contrast, symmetrical, regular, repetitive, and parallel patterns are not among the most prized works of art for their boring monotony. Instead of finding them in museums, we see these patterns in furniture upholstery or curtains not for contemplation but for soothing non-attention as a background. If we consider that the intensification of experience explains the riddle of the seduction of art, a colonoscopy is also an intense experience without being classified as a work of art. What we admire in art is the dexterity, the talent, the perfection accomplished, exactly as in sports. It is not the object, be it a painting or a football goal, but the performance of the play, poem, or game that we esteem.

Merlin Donald (2006) sees art as cognitive engineering to influence the minds of an audience. He proposes three stages of cognitive evolution: mimetic, mythological, and theoretical that overlap "as scaffolds" one over the other. It is strange to find a quasi-Hegelian style of speculative thought in an evolutionist and the choice of a mechanical metaphor such as "scaffolding,"[58] but more curious is attributing the theory of mimesis of art to Erich Auerbach rather than to Aristotle. He points out that the human species is unique in its ability to imitate. Why he does not consider cases such as of the chimpanzee Kanzi who was able to mimic words, or Imo's students that mimicked her washing potatoes and wheat? Nor does he take into account parrots who imitate sounds and birds that learn their dialects by imitation, as well as the polyglot warbler (*Acrocephalus palustris*) that is capable of imitating environmental sounds such as automobile motors. (Donald 2006, 19)

Ellen Dissannayake (1996) decides to track back artistic expression among social groups that, according to anthropologists, keep features of the evolutionary past of the human species, or the so-called "aboriginal," "native," or groups of hunters and gatherers (foragers) in general. What today is meant by "art," namely concerts, plays, paintings, or sculptures, is explored by researchers of art evolution among contemporary hunter-gatherers in order to assert that art is universal and an adaptive activity that results from a biological basis.

In addition to the ambiguity of the use of the terms "aesthetics" and "art," another vagueness infiltrates in her proposal, namely the notions of "adaptation" and "adaptive" clearly distinguished by Thornhill, since art can hardly be an adaptation, not being an organ or organism. Theories of adaptive behavior, however, imply a teleology that is alien to the evolutionary paradigm and to Darwin's theory (unless it refers to the aim of survival), so to assume that organisms live in order to adapt or develop adaptive activities (instead of understanding evolution as the process that retroactively explains present organisms as adaptations over millions of years) is inaccurate. As the metaphorical use of "language" invalidates Dutton's syllogism, in Dissanayake the metaphorical use of "art" in preliterate cultures taken literally operated the same wrongdoing. And it is a metaphorical use of "art" because, as pointed by Danto and Dickie, conventions and institutions are required to endorse, produce, and classify objects under that category, since it is not an ontological condition but a social convention. Aesthetics, on the contrary, is a universal condition of all live beings, and so it is meta or rather subanthropic. And so is poiesis evolving from praxis.

George Dickie (1974) argues that among other things, art is an institutional presentation of objects as candidates for aesthetic appreciation, where critics trained to perform such evaluation grant legitimacy to such objects and confer on them the quality or category of "art." Danto (1964, 1981) coined the term "artworld" within which art is produced and conferred such status as a result of experts' interpretation, classification, and indexing these objects as art. According to this institutional position, in effect, no animal art is possible, except through Komar and Melamid's humorous curating of the so-called elephantine art, exhibiting ad absurdum the artworld's bias, much like Duchamp did with the "Fountain" of R. Mutt in 1917 who re-contextualized a urinal in a museum and resemantized it as an artwork.[59]

Dissanayake (2007, 8) coins the term "artification" assuming that "art behavior or 'artification' is the overarching motivational system or adaptation composed of dancing, singing, decorating, carving—the various arts." But it's a fact that when we dance we do not try to perform "a dance" qua art but we simply enjoy ourselves in the trouble-free and sometimes ridiculous gesture of dancing around, as it happens at a wedding party or a sunny morning and consists only of moving around rhythmically. We dance for several reasons: for seduction (tango or danzón), for exhibition of social status (waltz or the tarantella), for its effects in the mind as the voodoo trance, for stimulating their team by cheerleaders, and only in the case of display of professional virtuosity in a modern or classic dance show can we speak of art. Moving the body rhythmically is also performed by the military during their marches, by priests in their processions or religious rituals, or in funerals when solemnly and rhythmically the relatives march with the coffin to the cemetery.

Smearing colors on a canvas, drawing doodles when bored at a meeting, or combining shapes and colors in our clothes, food, or furniture does not make us artists. Not all dancing is art though, in all cases, it is an aesthetic activity that involves rhythm and melody. The artification in any case is what Dissanayake made of the concept of aesthetics: by artifying it, she reifies it, hindering the biological context in its exploration. It is aesthetics, not art, that emerges directly from nature.

Homo ergaster, *neanderthal*, and *sapiens sapiens* cultivated various types of poetics by carving bifaces, singing, dancing and acting, body painting and shaping figures, or marking handprints on cave walls. This does not mean that our predecessors were artists, again, because art is a social convention requiring institutionalization and professionalization. Hence it is not possible to speak of "Paleolithic art," as there were no artistic conventions then, although "Paleolithic poetics" would be a more accurate term. What it means is that aesthesis and poiesis have a long evolutionary history, a million and a half years before art, predating even our own species.

The phenomenon of art, by its arbitrary character, its artificiality and strict dependence on context and conventions, is located at the opposite end of the evolution of our species. It operates by a constant feedback loop similar to what Fisher called a runaway process that today is consists of a totally self-absorbed and self-referential activity.

In the widely cited article "The spandrels of San Marco and the Panglossian Paradigm" Gould and Lewontin propose the idea of the spandrel (the triangular architectural element left over when a dome is supported over a column) as a metaphor for an existing item that fits another purpose. They coined the term of "exaptation," by contrast to "adaptation," for denoting a trait that evolved for certain purposes but is used in another context for a different function. (Gould and Lewontin 1979) Velez Caicedo (2008) follows the concept of exaptation to explain aesthetic activities of the homo artisticus through art theory. She proposes that an evolutionary perspective could be implicit in Gombrich's art theory, a very interesting route to take if armed by biological and evolutionary explanatory power.

In contrast to aesthesis which consists of an ongoing activity with different degrees of acuity, art is not an activity but a device or artifact. Art, as its name suggests, is artificial, the result of violating the natural tendency of things to make them into something different according to an intention or project. In Simmel's words: "The material products of culture—furniture and cultivated plants, works of art and machinery, tools and books—in which natural material is developed into forms which could never have been realized by their own energies, are products of our own desires and emotions, the result of ideas that utilize the available possibilities of objects." (Simmel 1997, 37)

Although it is possible to go from biology to culture for explaining art, the path is not straightforward: we have to cross a narrow doorway and solve a huge inequality of levels, greater than the twenty centimeters' difference on the Panama Canal to cross between the Pacific and the Atlantic Ocean. To trace the origins of art we have to focus our search on collecting rather than on creating (creation as in ongoing activity in different spheres, not only artistic, namely, technical, nutritional, cognitive, social, communicative). In other words, I am proposing here a collectionistic definition of art, not far in line from Danto, Dickie, and Binkley's position. While poetics refers to the making of objects that please aesthetically, art refers to their collection. Contemporary artists produce items for museums or private collectors, exhibited later in this institution. Hence to understand the evolution of artistic expression, this instinct for the display of collections is also present in the black wheatear bird (*Oenanthe leucura*) that collects a kilo and a half stones in the nest as a sign of fitness. (Zahavi and Zahavi 1997, 173; Moreno et al. 1994) Art as a collection of rare items is linked to its function for mating among different species (i.e., collecting food for seduction). It is remarkable to find, precisely in Pavlov, a comment that highlights the importance of the collection:

> If we consider collecting in all its variations, it is impossible not to be struck with the fact that on account of this passion, there are accumulated often completely trivial and worthless things, which represent absolutely no value from any point of view other than gratification of the propensity to collect. Notwithstanding the worthlessness of the goal, everyone is aware of the energy, the occasional unlimited self-sacrifice, with which the collector achieves his purpose. He may become a laughingstock, a butt of ridicule, a criminal, can suppress his fundamental needs, all for the sake of his collection. (Pavlov 1928)[60]

Research that takes a biological, evolutionary, and cognitive stand to explain art, such as empirical inquiries on aesthetic tastes and preferences and Zeki's school on neurocognitive research, are all making significant contributions to aesthetics. They help us out of the swamp of metaphysical speculation that could perhaps explain better such peculiar phenomena as a canvas with five strokes of painting auctioned for a million dollars, or that someone cans his excrement, puts it up for sale, and people are willing to buy it, as in the case of Piero Manzoni. Valuable clues for understanding artistic manifestations can be better found in the competition for status rather than in exploring sensibility, but since the issue focused on in this book is aesthetics, not art, I will not dwell further on this topic.

Without the tuning up to the edges of molecules (chemo-aesthesis) no detection of receptors to ligands can occur in the cell (cyto-aesthesis), on which plants (phyto-aesthesis) as well as animals (zoo-aesthesis) depend, sometimes displayed zoo-poetically by different species, including humans. It is precisely at this point, when a critical social mass assembles, that poetics exponentially soars to build its own unique universe through the development of artifice to mark hierarchies, differentiate roles, protect offspring, and evade mortality bursting with full force into the *Orbis tertius*.

NOTES

1. Jacobs et al. (2007); Monod quoted by Kauffman (1993, 25).
2. Quoted in Ricardo and Szostak (2009, 40).
3. http://www.evolutionnews.org/2013/07/gerald_joyce_de074891.html. Accessed June 23, 2014.
4. http://pewresearch.org/pubs/1107/polling–evolution–creationism. Accessed June 16, 2014. http://pewresearch.org/pubs/1276/science–survey3. Accessed June 16, 2014.
5. http://law2.umkc.edu/faculty/projects/ftrials/scopes/evolut.htm. Accessed June 20, 2014.
6. http://www.sciencemag.org/content/310/5749/787.1.full. Accessed June 20, 2014.
7. Ricardo and Szostak (2009) also support this possibility of protobiotic without genome.
8. Another newly labelled research field is "neuro-communication" that aims to explain in current terminology mind or "software level" and brain, or "hardware level" as a pair of coupled semiotic engines, i.e., computing devices for processing verbal and nonverbal signs. (Sebeok 1999, 90)
9. On bacterial communication cf. Heal and Parsons (2002); Greenberg (2003).

10. On the category of bonding or latching-on, which is the initial act for aesthetics, cf. Mandoki (2007). The whole of chapter 8 is dedicated to this concept.

11. In Barbieri (2008a, 126).

12. Nahon et al. (1995), quoted in Zahavi and Zahavi (1997, 104).

13. Biggers and Creed (1962) quoted in Trivers (1985, 122 fig. 6–18).

14. Darwin refers to twining plants like Pisum sativum, Echinocystis lobata, Bignonia capreolata, Eccremocarpus scaber, and with the leaf-climbers, Solanum jasminoides and various species of Clematis.

15. I.e., *Ceropegia*, *Sphærostema*, *Clerodendron*, *Wistaria*, *Stephania*, *Akebia*, and *Siphomeris*.

16. Sachs (1978) quoted by Hallé (2002, 211).

17. See also Robert and Roland (1989) quoted by Hallé (2002, 130).

18. On Phytosemiosis, consult also the canonical text by Martin Krampen (1981).

19. Cf. Mott (1990); Jang and Sheen (1997); Sheen, Zhou, and Jang (1999).

20. Quoted in King (1997, 57).

21. Part of this chapter was published in: Mandoki, Katya, "Zoo-aesthetics; a natural step after Darwin," in Timo Maran (ed.). *Semiotica: Journal of the International Association for Semiotic Studies / Revue de l'Association Internationale de Sémiotique*, *Special Issue in Zoosemiotics*, Volume 2014, Issue 198, Berlin: Mouton de Gruyter, 2014, pp. 61–91. I wish to thank the publishers for granting permission to reprint it.

22. See also http://infomotions.com/etexts/gutenberg/dirs/1/4/5/5/14558/14558.htm. Accessed June 16, 2014.

23. For Snowdon's paper see http://articles.sfgate.com/2009–09–09/news/17205342_1_tamarin–bird–songs–monkey. Accessed 5 May 2013.

24. Cf. Cronin (1993, chapter 9).

25. On different kinds of animal play see Bekoff and Byers (1998).

26. Andrew 1963b in Wilson (1980, 80).

27. I proposed the category of *peripatos* for game of enigma solving and applied Caillois's (1994) categories in Mandoki (2007, chapter 12); Caillois (1994).

28. As I have previously argued in Mandoki (2007a, chapter 14.)

29. (Trivers 1985. 414) quotes (Rohwer 1977).

30. Cf. (Thompson 1946).

31. Trivers in Dawkins (1976, xx).

32. I have argued extensively the Artistic matrix in Mandoki (2007, chapter 32).

33. For more bowerbirds' behaviour observations cf. Gerald Borgia (1985); Gerald Borgia (1995); Madden (2003); Borgia and Gore (1996); Gerald Borgia (1986); Borgia and Keagy (2006); Borgia and Collis (1989).

34. Cited in Darwin (1882, 413).

35. Morris 1962 reported in (Sebeok 1979, 37).

36. Described by Dr. Hartman in Darwin (1882, 282).

37. Bechstein (1840, 4) cited in Darwin (1882, 368).

38. Personal discussion during the 2011 Gatherings of Biosemiotics at Tartu.

39. This was the idea behind my "Parturition," 1978, Happening at the Art Gallery of Ontario, Canada. The public was led to believe I was actually giving birth on the Persian rugs at the gallery when in fact, I *was* giving birth, though not to a baby but to a sculpture. Since it was not a performance, but a happening, the public participated calling for "a doctor in the room" and worrying about my (difficult) situation. Was I cheating? Was I playing? Was I making a statement? Was I doing an artwork? All of them.

40. Yerkes 1943, cited in Trivers (1985, 158). For an example of this behavior in human children, see this video: https://www.facebook.com/photo.php?v=809236042434663&set=vb.148063885218552&type=2&theater Accessed July 6, 2014.

41. I will deal further with this concept in the fifth chapter, the display of excess.

42. See also Wickler (1966).

43. Cf. Butvill (2003); Norman and Hochberg (2005); Norman, Finn, and Tregenza (2001).

44. http://youtu.be/cNAAZ5Nt6pk, http://youtu.be/cJOZp2ZftCw http://youtu.be/GYMBIGTteWA. Accessed June 25, 2014.

45. Sachs 1937, cited in (Sebeok 1979, 17).
46. http://youtu.be/M8qprOG2LxY. Accessed June 25, 2014
47. I will deal with this question in chapter 5, in the section dedicated to Animal Semiotics and neo-Darwinism.
48. I developed the distinction between sensation, discernment, and regard in Mandoki (2007, chapter 14).
49. Emphasis in original.
50. MacLean cited in Pert (1999, 134).
51. The concept of integration of intelligence is similar to that proposed by Turner when he talks about blending or fusion. (Turner 2006)
52. Quoted by Darwin (1882, 584).
53. Sir S. W. Baker, "The Nile Tributaries," 1867, 121 and "The Albert N'yanza," 1866, vol. I, 218, cited in Darwin (1871, V2 339).
54. See also Grüsser et al.(1990) in Breidbach (2003).
55. (Wright 1994, 70) quotes (Symons 1979, 138–41).
56. Cf. Grammer et al. (2003).
57. On the concept of *languaging* cf. Maturana and Varela (1992, 234–235, 246).
58.. On this metaphor, see also Hoffmeyer Semiotic Scaffolding of living systems which is also strange considering it comes from a biosemiotic perspective. http://www.ut.ee/SOSE/tartu/suveseminar_08/hoffmeyer.pdf. Accessed June 20, 2014.
59. On an institutional approach to the definition and conceptualization of art cf. Danto (1964); Dickie (1974); Danto (1981); Binkley (1987); Mandoki (1991, 2007, chapter 32).
60. Quoted in Humphrey (1973, 436).

Chapter Four

Orbis Tertius

En la primera página y en una hoja de papel de seda que cubría una de las láminas en colores había estampado un óvalo azul con esta inscripción: Orbis Tertius.

—Borges[1]

For the human species, culture is nature. Wherever human groups dwell, culture is inevitably bred. Only in cases of feral people, isolated from their fellow humans, can we find individuals of our species devoid of culture. The origin of humanity is at the origin of culture because what culture cultivates is always communal life. We must therefore speak of "human cul(na)ture" because culture is the natural extension of our anatomy that expresses and transforms us as we transform it.

Unlike coral colonies that add new generations without altering their old configurations, cultural evolution can suddenly transform the cultural structure together with the biological environment that sustains it. This does not mean that culture is less dependent on previous substrates; on the contrary, it grows like a log that is covered with the young shoots upon both the living and the necrotic marrow of ancient customs. Whereas animals discard their excreta, culture, like plants, saves it, and does so in museums, libraries, and traditions.

Culture is the flora of *Orbis Tertius*, a third world whose big bang occurred 2.5 million years ago in Africa and whose traces still remain in Oldowan carved rocks. These artifacts, an index of a deliberate act to modify nature in a durable form, mark the precise moment in which homo habilis engendered homo culturalis. Carving the stone is also carving the brain and opening the amazing process of the mind-culture symbiosis.[2]

While coinciding at various ingredients like "learning," "imitation," "mind," "symbolism," or "language," there is no consensus on the exact margins for defining culture. A basic assumption held by neo-Darwinian evolutionary psychology is to understand culture as a result of an organ, the human brain, a morpho-psychological adaptation that evolved under the conditions and pressures of the Pleistocene. (Pinker 1997; Tooby and Cosmides 1992; Cosmides and Tooby 1992) However, the brain is a necessary but not sufficient constituent to address the issue of culture, for it leaves unexplained why then our Neanderthal relatives, whose brain was larger than their contemporaries Sapiens did not develop a more sophisticated culture. It would also need to explain the absence of elephantian, gorillian, or whalean cultures whose brain mass is much greater than ours.

This highlights the collective and social aspect of culture, for individuals of the three species mentioned above are solitary or live in small groups. However, creatures with a microscopic brain like ants, bees, or termites have a very complex social organization which allows us to already qualify it as a real culture regarding work division, home construction, efficient communication, and also imitation and language (i.e., the more experienced ants teaching routes to get provisions to their sisters).

If it is not the brain that explains culture, perhaps culture may explain the brain. Robert Trivers argues that the impressive development of hominids' brain volume during the Pleistocene may be due to the increasing complexity of social interactions, which required calibrating exchange relationships, altruism, reciprocity, punishment of abusive members, identification, trust, simulation, fault repair, loyalty, betrayal, and friendship, all of which require memory, classification, and learning. (Trivers 1971, 54) John Tooby and Leda Cosmides (1992) assume culture as an organization to care for and take care from others. For Peter Richerson and Robert Boyd (2005, 126) "A little scientific theorizing is necessary to convince us that the existence of human culture is a deep evolutionary mystery on a par with the origins of life itself." For them, "culture is as much a part of human biology as walking upright." (Richerson and Boyd 2005, 7) Other approaches that deal with culture from an evolutionary perspective are behavioral ecology, epidemiology of inner or outer representations (Sperber, 1996), contagion of memes in Dawkins (1976) or *culturgéns* for Wilson in *Consilience*, and as the fourth dimension of inheritance based on language by the human species (Avital, Jablonka, and Lamb, all quoted before).

The transmission of information occurs possibly from the *primus*, certainly throughout *secundus* by the *significans* web, and forcibly through the *tertius* by the *cogitans* network of crossing messages, exchange, and interaction.[3] In this sense the *tertius* is dependent on the *primus* for its material and on the *secundus* for its carriers and receivers. We cannot produce any events or artifacts in the *tertius* that are not determined by conditions of these two worlds. A text requires a code, a context, and an interpretative community in the *tertius* as much as it requires brain activity of its authors and readers in the *secundus*, and a physical support in stone for a stelae, or a parchment, paper, and ink in the *primus* to endure until another reader relives and gives continuity to this common semiotic vehicle. For this simple mismatch between transmitter and receiver, the *significans* web of *secundus* enters the temporal dimension that is projected into the future from the past allowing an exponential accumulation of experience. Culture recovers the process of relay of genetic, epigenetic, and behavior inheritance and catapults it to multiple potential beneficiaries. As Robert Wright (1994, 27) points out, since the main environment that humans adapted to has been the human environment "evolution of human beings has been to adapt to each other." This adaptation is precisely what we understand as culture.

Just as there is a narrowing or closure in the *orbis secundus* upon the *primus* to prevent that just any kind of molecules would form life, so entrance to the *orbis tertius* narrows to homo sapiens as the favorite heir of culture, with a few other much more modest and less conspicuous legatees. This favorite son of evolution carved his inheritance on Oldowan stone in the image and likeness of his imagination. It was the morning and afternoon of the seventh day of evolution, but if we count from the origin of the world, it was only one minute before midnight.

M

Mama, mom, moeder, mère, mutter, mother, Mae, мать, meme, маці, майка, mare, majka, מוטער, manman, matka, mor, ema, móðir, mháthair, אמא, matris, Mate, motina, omm, mor, mati, Mae, Mamu, mam. It seems no coincidence that in so many languages the word to denote the nourisher has an "m," as it imitates the baby's mouth ready to be breastfed.[4] For Steven Mithen (2005, chapters 9 and 15), "Hmmmmm" denotes the holistic, manipulative, multi-modal, musical, and mimetic origin of language. Based on Allison Wray's holistic view of protolanguage, Mithen proposes that Neanderthals communicated by song, rather than by language as a means of expressing and inducing emotion.[5] By contrast, he argues that the homo sapiens separated them, using music for expression of emotion and language for transmission of information. Vocalizations may have emerged from selective pressures during the Pleistocene's changing environments and thus required more intensive and precise information and communication on food supply, hunting planning, and mating, as well as for group cohesion enabling the development of language.

Dean Falk (2010) suggests that when human hominids lost hair, the origin of language could have occurred as a result of the need to maintain contact between mother and baby while leaving the offspring on the floor to use both hands for collecting food. Thus arises what he calls "motherese" or "baby talk," although it seems that the simplest noise would have sufficed for this end. Hens are pretty good at keeping track of their chicks without using hands or complicated language.

Whoever initiates communication is the one in greater need, and since the baby requires the mother's breast more than the mother's breast the baby (although the urge to breastfeed is really intense) language begins here. This process allows us to convey meanings that, with habit, become standardized by acoustic consistency to save energy in signaling for mutual benefit.

I suggest considering four concepts that can be useful for understanding processes of coordination that are crucial in the development of culture and the evolution from *significans* to *cogitans*:

1. The concept of *abstimmung* as intonation by von Uexküll for the constitution of collective orders. (T. von Uexküll et al. 1993)
2. The idea of "languaging" proposed by Maturana and Varela (1992) as recursive flows to guide behavior or coordinate and build consensus for action. This concept is complementary to the "language as work" of Rossi-Landi (1983), namely the physical, concrete and material task of language.
3. The Principle of Coordination (all discursive fragments led by a member of the communicative community to another member of the same serve to homogenize the internal structures of the members of that community).
4. The Principle of Cooperation (every piece of discourse that serves to homogenize the internal structures of the members of the community, tends to homogenize both systems), the latter two proposed by Parret (1993; 1995).[6]

Language and thought are dialogical (in Bakhtin's sense) and depend on a community since we always talk as though with someone, even if our interlocutor is imaginary or part of oneself in another mental role. Therefore *cogitans* being *co-agitare* or agitating with someone, involves interaction processes, all of which are products of mental dialogue. According to Norbert Wiley (1994), the semiotic self's internal communication (mentalese) operates through three temporal and linguistic levels in the scheme of "I-you-me." Wiley refers to the instance "I" which refers to the present, the "me" on whom it is spoken as identity in the past, and the "you" as the orienting instance towards a task or course of action to the future. To illustrate this scheme, at this very moment *I* am muttering the words that the *you* writes on the keyboard having the *me* as the conceptual context from which it is possible to develop ideas that the *I* communicates. By the dialogue with the *you*, *I* will perpetually integrate its actions that slide the *you* into the *me*, but never coincide except through the narrow evanescent *I*.

The image of the self that is co-agitated with another subject has erotic connotations, in addition to the *you* and *me* dialogue, because the brain, like our skin, is an organ that looks for contact and allows us feeling close to others through the mind. The *cogito* of the *I* with the *me* generate a narrative memory that maintains experiential consistency and a sense of identity. In these terms perhaps the Cartesian *sum* is *me* to the extent that the *cogitare* involves at least one other, whether it be *me* in the past, *you* of the future, the *du* in Buber, or you reader to whom I write. That first deliberate knock on the Oldowan rock occurred at a time when the *I* told the *you* how to hit the stone with an idea of what is possible that was kept in the memory by *me* 2.5 million years ago.

Besides these three temporal instances in Wiley's semiotics, internal communication is performed through four registers: scopic when we mentally visualize an image before we take action (the symmetrical handaxe while carving stone), lexic when linguistically articulating what we write or say (what Fodor referred to as "mentalese"), somatic when the body projects an action before or during the act, and acoustic when setting the emotional tone for interior and exterior language. Our mind has a tuner for volume.

If *cogitans* is co-agitation, there is no private language (in Wittgenstein's iconic phrase) as there is neither private knowledge nor private culture, because culture consists precisely on being a bridge among minds. Therefore, for the evolution of language and culture, a critical mass of partners was required for the homo linguisticus to emerge.

THE IDEALISTIC SLIME MOLD AND THE THIRD ORDER

Yeast can be amorous (described in chapter 3) and slime mold altruistic. *Dictyostelium discoideum*, commonly called slime mold, is a nucleated celled creature belonging to the taxonomic kingdom protista, living independently and engulfing bacteria by its pseudopodia all day long. When it detects food scarcity it emits a chemical signal that calls upon others of its kind to congregate and organize in elongated figures as pistils for producing spores that migrate to more benevolent environments. This species is remarkable for its sudden transformation from a single-celled independent, solitary organism into a pluricellular cooperative collectivity where the basic elements that make up all social order suddenly emerge. We find here: 1) semiosis (chemical communication), 2) division of labor in building the pistil, 3) agreement in gathering for the same purpose, 4) altruism, as only cells at the tips of the pistils form spores to reproduce their genes, not in the base. Yet, since mold is made of identical twins who share the same DNA, this altruism is explained by William Hamilton's concept of kin selection, i.e., cooperating with genetically related individuals to promote their own genes.

(W. Hamilton 1963) In these caste systems of ants, termites, and bees, individual agents generate collective behaviors at a more complex scale. How the division of labor and roles occurs among these species we do not fully understand, but it is a fact that a chemio-biological process takes place, for example, in the honeybee queen who secretes 9-keto-2-decenoic acid to inhibit the development of her sisters' uteruses and turn them into sterile workers at her service (or the opposite: the queen is at the service of worker bees). (Wilson 1980, 99)

We can observe collective sensitivity at work in swarms by the coordination of flying or in the immune system's reaction to viral infections by the division of labor between lymphocytes and macrophages. It does not seem merely a juxtaposition of individual behaviors but the emergence of a third order *umwelt* through which a group is perceived as a whole by a different kind of sensitivity, namely the aesthesis of regard for others and collective semiosis. Such regard or communal sensibility involves a dynamic of third order self-organization over the role each individual must take in the system to reproduce itself. Hence, it is necessary to distinguish between aggregation and society, and between social organisms and cultural organisms, since culture and society are often, by mistake, used interchangeably. I will thus understand society as structured, differentiated, interactive, and hierarchized collectivity of the same species, and culture as the intergenerational result of this group's organization stored in noncorporeal formats.

Among non-anthropocentric definitions of social groups, Varela and Maturana propose the concept of third-order units (the first are unicellular, second are multicellular individual organisms, and the third multi-individual groups) including the colonies of *Hymenoptera* and termites. (Maturana and Varela 1992) Another definition comes from sociobiology: "a group of individuals belonging to the same species and organized in a cooperative manner." (Wilson 1980, 7) Wilson proposes the following categories for analyzing collectivities: a *colony* is, "in strict biological sense . . . a highly integrated society of organisms," united either by physical union of bodies, caste division by specialization, or both; a *population* is a series of organisms belonging to the same species occupying a well defined area simultaneously; and *group* similar to the previous but also interact more frequently with other organisms of the same species. He adds the key categories of sociality in various species such as *communication*, *coordination*, *hierarchy*, and *regulation* and puts forward the following qualities of sociality: size, demographics, cohesion, pattern and amount of connectivity, permeability, compartmentalization, role differentiation, integration of behavior, information flow, and time devoted to social behavior.

John Maynard Smith and Eörs Szathmáry propose the emergence of these new levels of complexity in eight stages: 1) from replicating molecules to populations of molecular compartments (protocells); 2) from independent replicators (genes) to chromosomes; 3) from RNA as gene or information vehicle and enzyme to information vehicle such as DNA and protein enzymes; 4) from prokaryotes to eukaryotic cells; 5) from asexual clones to sexual populations; 6) from unitary eukaryotic cells to multicellular organisms with differentiated cells (plants, fungi, and animals); 7) from solitary individuals to colonies and social groups with castes that do not reproduce; and 8) from primate societies to human societies with languages that allow memes. (Smith and Szathmáry 1997, 6) All these changes, it must be noted, depend on the way semiosis is stored, interpreted, and transmitted from *significans* to *cogitans* at *secundus* and *tertius*.

We can see from this perspective that an aggregation process was initiated in the *primus* by the assembly of quarks in neutrons and protons that are bonded in atoms and molecules, to carbon molecular chains that evolved replicating nucleotides inaugurating *secundus*, that formed chromosomes to make peptides to build proteins enabling prokaryotic cells which in symbiogenesis generated eukaryotic cells gathering in multicellular organisms that share the same DNA. Individuals mate and procreate, become families, groups, tribes, colonies, populations that form civilizations which cooperate (or violently compete) for their reproduction (or destruction), and share (or fight) a culture.

> The dawn of the social thus is not a link in any chain, not a step in a path, but a leap to another plane. It may be likened to the first occurrence of life in the hitherto lifeless universe, the hour when that one of infinite chemical combinations took place which put the organic into existence, and made it that from that moment on there should be two worlds in place of one. (Kroeber 1917)

ACHILLES AND THE CULTURE

$$1 = \frac{1}{2} + \frac{1}{4} + \frac{1}{8} + \frac{1}{16} + \frac{1}{32} + \dots$$

When the tortoise challenged Achilles to a race on the condition of giving her an advantage at the start, she knew Achilles could never reach her. To do so, he would first need to cover the distance that the tortoise advanced, and in order to do that, he would have to reach half of it, and before that its half, and so on *ad infinitum* according to the well known Zeno's paradox. Something similar has happened to the definition of culture, because we still do not know enough of the mental world of animals or zoo-*cogitans* (I imagine Descartes revolving in his grave by the aberration of this concept) to eliminate the possibility of cultural transmission in different species. What we do know is that very primitive species can have a very sophisticated power of induction, as observed in crabs by Gardner quoted by Darwin:

> [A] trustworthy naturalist, Mr. Gardner, whilst watching a shore-crab (Gelasimus) making its burrow, threw some shells towards the hole. One rolled in, and three other shells remained within a few inches of the mouth. In about five minutes the crab brought out the shell which had fallen in, and carried it away to a distance of a foot; it then saw the three other shells lying near, and evidently thinking that they might likewise roll in, carried them to the spot where it had laid the first. It would, I think, be difficult to distinguish this act from one performed by man by the aid of reason. (Darwin 1882, 270)

To establish criteria for characterizing culture without neglecting the importance of animal culture, we could list several defining attributes of culture:

1. Learning and transmission of behavior (words by bonobo Kanzi or chimpanzees Washoe and Nim Chimsky, local dialects in birds or the domestication of dogs, cats, parrots, and tits). (Hess 2008; Gardner and Gardner 1969)
2. Division of labor (social insects like ants, termites, and bees).
3. Playing (birds, monkeys, cats). (Bekoff and Byers 1998)
4. Creativity and inventiveness (Japanese female chimpanzee Imo who discovers to wash potatoes or wheat in water to remove the sand). (Kawai 1965)
5. Social behavior as the herd instinct, territoriality, or cooperation (groups of mammals and fish).
6. Rituals and conventions (mating, flight rhythms, ritual fights between moose).
7. Construction of artifacts (nests built by birds, snakes, fish, or mice; beaver dams; termite combs and nests; spider webs).
8. Language (language in humans, zoo-semiosis in other species as analogic alarm calls by birds and monkeys or digital bee dance).
9. Imitation (guppy fish, monkeys).
10. Altruistic behavior (dolphins rescuing peers in penury, adoptive wolves, elephants cooperating and mourning the dead).

Each of the concepts listed above is a necessary condition for culture, but none is sufficient. If we define culture by the ability of learning, then we have to acknowledge the culture of ducks, geese, and swans who learn travel routes from their elders; of monkeys who learn from their mother to dig roots or wash potatoes. The routes of the deer are transmitted through generations; turtles learn to migrate and leave their eggs in specific places (like Mazunte), and birds learn complex dialects from their group. (Wilson 1980, 87) Already at the level of ants (*Temnothorax albipennis*) we may find learning through a mechanism called "tandem running" where an experienced worker guides others by imitation to food sources. These cases cover at least two of the conditions that define culture, namely, imitation and learning.

Animal traditions, like humans', are caused by the transmission of information through languages with partly arbitrary elements and indexes of the emitter's status. Thus we can see the six functions of language proposed by Jakobson in action through several animal species (Jakobson 1960): 1) the *expressive* function in birds and mammals' alarm cries; 2) the *referential* function in the bee dance to indicate the location and quality of pollen; 3) the *conative* function in the growl of an alpha male to force a weaker individual to yield and subordinate (other cases are cricket stridulation to seduce the female, as well as the chicks' chirping asking for food or the screaming tantrum among monkeys to manipulate the mother); 4) The *phatic* function is used by the hen in monitoring chicks by chirping; 5) the *metalinguistic* function is present in the gesture indicating "this is a game" by a playing monkey (as detected by Bateson, mentioned before); 6) and lastly the *aesthetic* function in the bearing of a bird of paradise that not only displays its tail but displays itself displaying it, as we saw in the section on zoo-poetics.

Emerging signs of cultural behavior are present among birds' transmission of local dialects as in the white crowned sparrow *Zonotrichia leucophrys*. Each group member maintains an iconic acoustic song with that acquired trait that may be relatively short-lived, as it depends on the living voice and repetition, but involves learning of a conventional dialect. Certain behaviors among animals are inherited, like dogs showing teeth to threaten attack, while others like waving a paw to ask for food are learned. There are also cases of proto-cultural transmission in collecting food as among black rats (*Rattus rattus*) learning from their mothers how to take seeds from pine spiral cones, a case reported by Terkel (1996).

For Whiten "cultural behavior is one that is transmitted repeatedly through social and observational learning until it becomes a feature at the population level." (Whiten et al. 1999, 682) There is also individual learning: Sebeok tells of weaver birds that improve the quality of their nests, making them more symmetrical, clean, and elegant than in the previous season. (Sebeok 1979, 31)

Imo is a paradigmatic case of animal learning and invention. She is a chimpanzee living in Koshima Island, Japan, who according to observations by Kawamura and Kawai, invented washing potatoes to clean off sand. Moreover, she transmitted her invention to her pupils, who learned to wash potatoes and wheat. (Kawai 1965; Kawamura 1959) In Gombe National Park, Tanzania chimpanzees remove leaves from a branch to use as it as a tool in termite fishing, thus exemplifying the ability to imitate and learn in various animal species. Kanzi, a bonobo, spontaneously learned English while humans tried to teach lexigrams to his mother using gestures and vocalizations, to the extent that he could take action with the skill almost of a two and a half year old by obeying phrases like "go out and bring carrots" or "spill the coke in lemonade." (Jablonka and Lamb 2005, 338)

Whiten and Goodall made a comparative summary of sixty-five chimpanzee behaviors through observations spanning 151 years at seven sites in Africa and conclude that there is definitive evidence of imitation and individual learning in that species. They drew a typology on the following terms: 1) behaviors that depend on the environment, 2) absent in all behaviors, and 3) that are repeated in all groups without variants (indicating that they may be instinctive), which gives a total of thirty-nine behaviors that vary from one region to another. These behaviors represent local learning and therefore we may adequately call a regional culture. They involve behaviors such as grooming, use of tools, courtship.

> We find that 39 different behavior patterns, including tool use, grooming and courtship behaviors are customary or usual in some communities, but are absent in others where ecological explanations have been ruled out. Among species of mammals and birds, cultural variation has been previously identified only for individual behavior patterns, such as the local dialects in bird songs. The wide and multiple variations now documented in the case of chimpanzees are unparalleled. Moreover, the combined repertoire of these patterns in each chimpanzee community itself has a highly distinctive character, a phenomenon characteristic of human cultures but has not been recognized previously in nonhuman species. (Whiten et al. 1999)

Local differences in behavior, according to Galef, are not sufficient to assume the existence of instruction or learning that results in the animal culture. He explains these behaviors as a result not of imitation but of the frequent presence of stimuli in the context, as finding sticks near a termite nest and using them to fish through holes.

> Through evolutionary time, a population of our direct ancestors may have become increasingly likely to express culture as a consequence of selection for increased sophistication in intellectual processes that served other functions and only later came to play a role in cultural transmission. After the thresholds of culture had been crossed, once our hypothetical ancestral hominid was able to imitate or to teach, then selection for increased ability to participate in culture (for increased "docility," in Simon's [1990] terminology) would have led to elaboration of intellectual processes supporting cultural transmission. In this model, continuity would be found in the evolution of psychological processes necessary for culture, not in culture itself. (Galef 1992, 173)

He accepts that there may be animal traditions, but not animal culture, and emphasizes the difference between homology and analogy with human culture. Yet having such "psychological processes necessary for culture," how to explain that there could be no culture in any of these species?

Lee Alan Dugatkin (1996) assumes that imitation may exist in animal species as in the case of sexual selection in guppy fish. The guppy females or

Poecilia reticulata of Paria River in Trinidad have an inherited preference for selecting the more intensely orange colored. This instinct of attraction to color (or aesthetic bias) is explained as an adaptation, because the most intensely colored are usually the bravest against the predators. Dugatkin notes also that some females choose a less colored male when copying or imitating the choice of other females, provided that the male is more assertive and only if the difference in color was not too great, otherwise it is reversed.

> Three treatments were undertaken in which males differed by an average of 12 percent, 24 percent, or 40 percent of the total orange body color. In all cases, observer females viewed a model female prefer the less colorful male. When males differed by 12 percent or 24 percent, observer females preferred the less colorful male and thus copied the mate choice of others, despite a strong heritable preference for orange body color in males. When males differed by 40 percent orange body color, however, observer females preferred the more colorful male and did not copy the mate choice of the other female. In this system, then, imitation can "override" genetic preferences when the difference between orange body color in males is small or moderate, but genetic factors block out imitation effects when the difference in orange body color in males is large. This experiment provides the first attempt to experimentally examine the relative strength of cultural and genetic preferences for a particular trait and suggests that these two factors moderate one another in shaping social behavior. (Dugatkin 1996)

Boyd and Richerson propose a concept of culture from a systemic perspective of genetic and cultural co-evolution. "By 'culture' we mean the transmission from one generation to the next, via teaching and imitation, of knowledge, values and other factors that influence behavior." (Boyd and Richerson 1985, 2) The teaching and imitation mechanisms may include nonhuman behaviors as birds' dialects, Imo's students washing potatoes, and probably others unstudied in many other animal species. Yet, in referring to "values" Boyd and Richerson would seem to exclude all non-human species. However, if we understand the values as positive or negative environmental stimuli that guide decision and action in a horizon of bifurcations, the concept can be extended beyond its anthropocentric sense. I mentioned before the cases of altruistic behavior, as in Tursiops dolphins attempting to raise a newborn above the water to help him breathe and save his life. There are cases of adoption of orphans by wolves or chacma papios as well as birds who adopt orphans or abandoned chicks even of other species and care for adults who have been blinded.

Consider that in all functional cycles, values are always at stake relative to the direction of the perceptor towards the effector, as sunlight for plants, pollen to the bee. There are also different types of values concerning the various dimensions and social institutions that frame and decide an action in the *tertius*: currency in economy, words in language, loyalty and cooperation in groups, skill and talent in art, rigor and excellence in science. These values extend through a plane of relations of opposition, contrast, or complement relative to others in the system (i.e., differential values of signifiers in Saussure's linguistic paradigm or of commodities in the market in Marx's political economy paradigm).

Twenty years later the authors refine this definition and propose that "Culture is information capable of affecting individuals' behavior which they acquire from other members of their species through teaching, imitation, and other forms of social transmission." (Richerson and Boyd 2005, 5) Values disappeared, and instead the concept proposed is information as "any mental state, conscious or not, that is acquired or modified by social learning and affects behavior." Again this definition encompasses animal species since a mental state is not unique to the human species, reminding us of Gregory Bateson's classic definition of information as "a difference that makes a difference."

Eytan Avital and Eva Jablonka define animal culture in the following terms:

> We are going to use the term "culture" for "a set of behaviour patterns or products of animal activities that are socially transmitted in an animal lineage, group, or population." When referring to the symbolic aspects of human culture, we will say explicitly "symbolic culture." Essentially, what we are suggesting in this definition is that, even when symbols are not involved, if there is a socially transmitted package of behaviors or products of animal activities, it is legitimate to talk about it as "culture." (Avital and Jablonka 2000, 22)

Jablonka and Lamb propose the existence of animal culture as they communicate across generations and emphasize the active and reconstructive factor of the information acquisition process.

> [C]ulture is the system of socially transmitted patterns of behavior, preferences, and products of animal activities that characterize a group of social animals. Cultural evolution was described as a change, through time, in the nature and frequency of these socially transmitted preferences, patterns, or products of behavior in a population. Nonhuman animals transmit behavioral information in a variety of ways: often it is through vocal signs, as in the communication systems of birds and whales; in other cases it may involve a complex combination of vocal, visual, tactile, and olfactory signs. When communicated across generations, these animal signs may form a culture. (Jablonka and Lamb 2005, 205)

According to the authors, we can explain cultural transmission among animal species when a successful discovery becomes habitual and instinctive until it eventually affects genetic selection. Culture would begin in an overlap on biology that unravels as we get closer to what they define as "symbolic culture" denoting human language.

From Peircean semiotics, besides the conventional symbols of language there are also other processes of meaning in iconic signs such as diagrams or pictures and indexical signs produced and interpreted as traces. All animals use indexical signs like sniffing dog urine or ants emitting and interpreting pheromones, male monkeys decoding estrus indexes, cells detecting ligands, antibodies detecting antigens.

Animals are capable of lying, performing rituals, playing, working, imposing territorial barriers, indicating they are joking, learning and teaching techniques. They can even send each other altruistic alarm calls that endanger themselves or commit suicide by biting a creature that threatens her partners, as does the bee. Several species obey rules, communicate, admire, console, or take care of the orphans and the destitute (as do beavers, bower birds, bees, ants, and spiders). Unlike humans, among whom, according to FAO's 2014 statistics, more than three million children die of hunger every year, the culture of the *Hymenoptera* ensures there is food for all its members and covers their basic needs as shelter and company.

As our discernment of animal behavior turns increasingly finer and subtler, the fuzzier the borders become, something akin to measuring the coast of England described by Benoit Mandelbrot: the smaller the measuring scale the greater the measurement becomes. Therefore, whenever we want to draw a definitive boundary between humans and animals we become Achilles trying to reach the tortoise, always entangled by intermediate points.

LITIGATION AMONG HEIRS

First sell me your birthright.
—Genesis 25:31

We have in our possession a testament written in code, transcribed more than 3.5 giga years ago, inherited by evolution, signed to our name and sealed in our genes. About this inheritance, neo-Darwinism has declared itself a self-appointed executor claiming to control by the gene all bodies, minds, social relations, and cultures. Although genetic heredity is undeniably the firstborn (no one would dispute), it is not the only heir. Aristotle, William Harvey, Karl von Baer, Ernst Haeckel, James Mark Baldwin, and Jan Piaget have fought for the recognition of an additional heritage that is not predetermined before birth and keeps transforming the heir through its growth. Evidence on this epigenetic legacy has increased in recent years, forcing the firstborn selfish gene's genocentric petulance to accept that there are things that genes cannot do nor explain. Five hundred and thirty million years ago this other legacy was formally drafted and given by the mother cell that voluntarily provides her daughters with instructions for differentiated reproduction. This second will dictates who will be a liver cell, a blood cell, a skin cell, or an egg cell, demanding that each configure herself for serving this task. Such inheritance is transferred by epigenesis and is exclusive to multicellular creatures through a process called "methylation" that selectively activates or numbs genes so that cells transmit particular information to their heirs.

When creatures evolve a more complex nervous system, they receive a third heritage, which is a behavioral legacy. This bequest was observed and recognized by Darwin in sexual behavior as another form of selection and transmission of traits. By behavioral transmission, creatures learn and pass on habits of their species: birds receive a legacy in song repertoires, the London tits the skill of opening milk bottles and even eating the cream, some rats learn the technique of spiral peeling seeds from pinecones. Baldwin, the author of *Organic Selection*, speaks of the continuity of the social environment in animals to adjust the ways and habits of their community. He defines social heredity as "the handling on of that which has been acquired by preceding generations; and I have used the term 'social heredity' for the accommodation of the individuals of each generation to the social environment." (Baldwin 1897, 635–36) What today is known as the "Baldwin effect" is the idea that through a learned response to the changing environment a genetic basis evolves. The notary of this heritage was Lamarck and this will has been in heated litigation for over a century.

A fourth legacy has bequeathed us chairs to sit on, books in our library, cuckoo clocks, words, music, balloons. To be entitled to this inheritance we require, as above, the nervous system with the additional prerequisite of very flexible neuronal connections in the cerebral cortex.

Jablonka and Lamb proposed four types of inheritance (genetic, epigenetic, behavioral, and linguistic) to question the genetic hegemony of neo-Darwinists. Their concept helps explain the diversification of organisms among which only humans inherit all four dimensions. The authors then defined culture as "a pattern of behavior, preferences, and products of socially transmitted animal activities that characterize a group of social animals. Cultural evolution was described as a change over time in the nature and frequency of these preferences, patterns, or products of socially transmitted behavior in a population." (Jablonka and Lamb 2005, 205)

For Boyd and Richerson, culture is a more agile and efficient adaptation to changing environments than genetic adaptation, which would never have evolved unless it could do what genes cannot. They emphasize that genetic inheritance occurs before birth and is determined at the time of fertilization, while cultural heritage begins after birth and has no pre-established limits, only statistical probability that could lead to a variety of directions. "Culture is an evolving product of populations of human brains, brains that have been shaped by natural selection to learn and manage culture. Culture-making brains are the product of more that two million years of more or less gradual increases in brain sizes and cultural complexity." (Richerson and Boyd 2005, 7)

To sum up, most anthropologists, like Maurice Bloch (1991) assume that only humans are capable of culture, as he equates culture with language. Jablonka and Lamb also circumscribe culture to human language or what they call "symbolic transmission." Boyd and Richerson are more encompassing as they consider information equivalent to culture, although, as we have seen in chapter 2, information occurs already at the level of genetic coding if not earlier and across the whole biosphere.

It is therefore important to note that in addition to these four qualitative leaps in evolution, namely genetics to epigenetics, to learning, and to language, there is a perfusion of signs that underlies them all along a semiosic spectrum by the *significans* network through the *Orbis secundus* and from which the genetic code to bacterial, animal, and human forms of communication evolve.

EVOLUTION OF EVOLUTION

We do not yet know whether culture is subject to natural selection processes, or whether cultural selection is a different form of natural selection. Perhaps there are different types of selection in *Orbis primus*, *secundus*, and *tertius* and different mechanisms and processes, objects and devices operating in evolution. We should also pay attention to the phenomenon of emergence of new forms, whether it is a case of random mutations only or deliberate planning and what factors come into play. The crucial question comes up: Does evolution evolve? I tentatively propose ten factors that may allow understanding biocultural evolution, examine their relationships and perhaps contribute to the answer of this question:

Table of Ten Evolutionary Factors

1. **Selection Orbs**: *primus*, *secundus* or *tertius*; their molecular, genetic, behavioral, and cultural breeding ground.
2. **Selection objects**: particles, atoms, molecules, genes, chromosomes, cells, organisms, individuals, groups, memes (fads, concepts, ideas, languages, traditions, theories, texts, habits, hobbies, prejudices).
3. **Selection devices**: ribosomes, chromatin marks, parents, teachers, opinion leaders, publishers, patrons, propagandists.
4. **Selection mechanisms**: molecular, natural (drifting), cellular, sexual, behavioral, cultural.

5. **Transmission devices**: DNA molecules, gametes, viruses, cells, images, stelae, scrolls, buildings, libraries, data banks, money banks.
6. **Transmission mechanisms**: mitosis, meiosis, genetic replication, epigenetic transmission, behavior copying, linguistic performativity, ritualization, mechanical reproduction, memetic contagion, schooling, imitation, imposition, indoctrination.
7. **Selection biases**: inertial (habit), entropic (least effort), direct (content) and indirect (frequency-rarity, status, aesthetic, political, technical).
8. **Emergence of variations**: mutations, inventions, guided variations, hybridizations, misinterpretation, deliberate distortion.
9. **Evolutionary forces**: a) pressures and resource shortages or economy of excess (pressure to dissipate), b) drives as the survival instinct, breeding, dominance, and territoriality. Pressures (a) push the body from the outside and the drives (b) from the inside.
10. **Co-evolutionary crossings**: genetic-behavioral, genetic-epigenetic, genetic-cultural, behavioral-cultural, epigenetic-cultural, and behavioral-epigenetic.

Orbs of Selection

In this context we can understand better the significance of the distinction between *Orbis primus*, *secundus*, and *tertius*. We do not know enough of the *Orbis primus* developments to explain why our universe evolved in the direction it did, why the quark and antiquark decreased dramatically to the union of three quarks into protons and neutrons, which in turn became hydrogen, helium, deuterium, and lithium nuclei and, after three minutes, matter and radiation were coupled for 300,000 years until the formation of organic molecules and compounds 3.7 billion years ago. (Hawking 1996)

If indeed a physical selection mechanism operates in the *primus* towards more stable configurations, to conclude that natural selection operates on the *secundus* and cultural selection on the *tertius* would need to be tested by the above factors that seem to evolve in crisscross rather than parallel or analogous patterns. It has only been verified that each new type of evolution stems from the previous, so that cultural evolution depends on the behavioral and the latter on the organism's morphology, which in turn depends on its genetics and on the laws structuring matter (as proposed by D'Arcy Thompson 1946).

Objects of Selection

Cronin (1993) describes the heated debate about the real objects of selection in the *secundus*: genes, individuals, groups, species, etc. Consider also particles, molecules, or types of energy in the *primus* and habits, memes, or civilizations in the *tertius*. Neo-Darwinism insists that the only object of selection is the gene and all other phenotypic components are devices under its service. Altruistic behavior in the Hymenoptera as ants and bees unable to reproduce and directly in service of the queen was a challenge for the selfish gene theory until Hamilton (1963) elegantly solved the puzzle with the concept of *inclusive fitness* or selection of same family genes. With a mathematical model he proves the existence of kin selection control exercised by the genes, since the bee workers are twin sisters to the queen so her offspring are genetically equal to their potential children.

At an epigenetic level, objects of selection are chromosomes that by methylation or activation define through chromatin marks the characteristics in the development of the cells according to their location. At the behavior level, objects of selection can be bird songs, mating rituals, role modeling, hunting, and oviposition behavior of certain species. At a cultural level, objects of selection can be fashions, languages, texts, idols, rituals, customs, canonical authors, famous artists.

Dawkins named "meme" the minimal unit of cultural replication equivalent to a gene in biological replication, but it seems to operate by a mechanism more akin to the viral contagion than to genetic replication, as he states that it invades brains "jumping" from one to another as hosts for meme propagation. (Dawkins 1976, 192–206) In order to propagate, however, memes must meet the same three requirements as genes: copying-fidelity, fecundity or replicability, and longevity or duration.

One of the problems that arise with the concept of meme (as in "culturgen" in Wilson) is it implies the idea of "minimal units of culture" that transmit it, yet Dawkins never explains what the limits or criteria for these units could be. These minimal memetic units have the same difficulty faced in the sixties and seventies by semiotics in their pursuit of "minimal units of meaning" such as the "sememe" or the "viseme" that freeze the living process of signification into a forced atomization. Levi Strauss questioned this problem regarding traditional semiology and proposed instead a systemic approach to relations and differences (although in the end he proposed the concept of "mytheme" as the unit of analysis of the myth, which, however, is not a minimal unit).

A very successful selection object in the *tertius* is the Pentateuch. The reasons for its widespread replication for nearly three thousand years depended originally on the tribal and ethnic matricial organization configured by the replication of the oral and written text and historically bonded to it and by it. It is a complex adaptation response to a deep emotional and mental need for narratives that convey meaning to the world and cohesion to the community.

Perhaps learnability could be considered as an object of selection which, as pointed out by Henry Brighton, allowed the emergence of culture under three principles: 1) the assumption of the innate or biological basis of learning (although it is not a specific capacity for language only), 2) the principle of situatedness where language universals cannot be explained only by the cognitive basis of language, and 3) the role of independence that makes clear that any of the functional properties of language are not determinants of the language structure. (Brighton, Kirby, and Smith 2005) I think that in any case such learnability is present already in the crab described by Darwin, when it removed the balls from the proximity to the hole after seeing one rolling in. (Darwin 1882, 270)

This selection object may vary in different degrees and be present in individuals of various species, such as chimpanzee Imo and her female students, the mockingbird and his great ability to imitate sounds, or the mimicus octopus who observes and imitates the form and color of other creatures.

In mate selection consisting of selecting genes for the next generation, both selection criteria are at play: biological (attractive body odor, body proportions, intensity of gaze, hair shine, shape of teeth, stature, complexion) and cultural (social status, education, lifestyle, learnability, charm, common cultural niches, artistic preferences, use of leisure time, wealth, elegance or vulgarity).

Selection Devices

Among selection devices in the *primus* we can consider laws of morphological conformation as studied by René Thom and D'Arcy Thompson. Such devices in the *secundus* correspond to ribosomes that transmit genetic code sequences for constructing new nucleotide chains to make cell proteins using amino acids. They also perform gene selection, for example in the mixture of cell chromosomes during meiosis. At the cellular level there are other selection devices common to bacteria, fungi, plants, invertebrates, and vertebrates through "chromatin marks" (components of chromosomes) that regulate traits transmitted to daughter cells by epigenetic methylation mechanism. On a macroscopic organism, sexual selection devices are phenotypic from fitness signaling to displays of dominant figures in some species as the alpha male leader of a group of mammals.

In the *tertius* figures that dominate certain cultural territories and institutions act as selection devices by the power of decision to propagate or not certain memes, namely editors, filmmakers, human resource managers of companies, research leaders, groups that control political parties, teachers, and so on. Under capitalist production and mass consumption, the selection device par excellence is rating.

Timo Maran proposes the concept of "semiotic selection" referring to animals' perception processes, among which he includes sexual selection as one type, since semiosis is a condition for sexual selection. (Maran 2009, 489)

Frédéric Kaplan (2000) emphasizes the concept of *semiotic schemata* from a mechanism he calls "cultural selectionism" to investigate the dynamics of replication of words in a given culture. He claims that words, like genes, are replicators competing to colonize our brains and that languages, as bodies, can be seen in competition with each other. Kaplan requires a selection device that is the individual and the objects of selection which are semiotic schemes, with noise as a pressure factor for the selection of the most resistant to distortion.

In the third *orbis*, we can consider Paul Grice's conversational maxims (avoid obscurity, ambiguity, be brief and orderly) as selection devices in receiving messages by quality, quantity, relevance, and manner. (Grice 1996, 125) We select messages that we assume are true, concise, clear, and relevant to the recipient or emitter and we discard saturated, redundant, semantically false, or informationally poor messages. Unfortunately, however, transmission of false messages is widespread, rapidly propagating even for generations and in various locations, such as horoscopes, myths taken literally, or libels and slurs against minorities, as well as subhuman and cruel traditions like ablation. Various factors come into play such as the discursive style, the channeling of interests promoted by a social group, the status of the issuer, the political situation, the chances of economic success and social coverage, and especially its emotional impact. Such impact depends on psychological configurations that respond to perception biases (which we will revisit in the last section of this chapter).

Agner Fog (1999, 95) proposes the idea of selection of the most suitable cultures, where cultures are like organisms that spread at the expense of others, and are subject to a selection process. He distinguishes between "selection k" or *kaliptical* and "selection r" or *royal* as cultural selection processes dominated by external or intergroup conflict and expansion process in space and time. The example he uses to illustrate his point is the USSR as selection *r* and USA as *k*.

A test case for the relevance of the concept of selection in the *tertius* is the Spanish invasion of Mesoamerica. It is difficult to decide here whether it was cultural or natural selection of cultural elements that were more replicable than the original Mesoamerican, e.g., Catholic ritualization and a more complex, efficient, and orderly hierarchical organization. European civilization conquered and destroyed—red in tooth and claw—the local civilization and cultural systems among the Aztec, Maya, and Inca when they burned their codices, imposed their own rituals and languages, and prohibited the local. Spanish conquistadors' technological devices were more lethal (firearms), their semiological devices more efficient for recording and transmitting information (European hybrid language and alphabetic writing), the social devices richer and much wider for geographical reasons with a very vast web of economic relations geared to the capitalist mode of production and colonial expansion. We cannot forget the smallpox biological device, besides the climate and geographic determinants in the *primus* that Diamond (1999) so clearly detailed and extensively explained.

If the destruction of Mesoamerican civilization could be considered as a case of natural selection in the *tertius*, the probable selection devices at work were: information about the context, the ferocity of the agents, their social capital in networks, group coordination and persistence, endurance and strength of their defense and attack teams, the abundance of funding. All according to the same pragmatic criteria of the *secundus*: an organism that receives more information and better protection is most lethal than the isolated organism whose defensive or offensive weapons are consequently less efficient.

Selection Mechanisms

Darwin and Wallace both hit upon the mechanism to explain evolution of species by natural selection in the survival and reproduction of individuals over many generations, now translated to the classic formula of random variation and selective retention. A century later, Rosalind Franklin (prematurely dead at 37), Watson, and Crick, among others, helped unravel the mechanical transmission of heredity at the molecular level in replicating DNA. Since then we are able to understand the unfolding dynamics of the double helix and the mediation of ribosomal RNA in its messenger, transmitter, and the nucleolus patterns.

At the next epigenetic level, the selection mechanism operates by methylation blocking transcription of certain DNA bases to activate others (because the acetyl groups have a looser chromatin structure than methyl groups). An index for cancer cells, for example, is the acceleration of DNA methylation patterns, which can be seen in error transcriptions by interaction between mutations and epimutations, or interactions between genetic and epigenetic levels. (Jablonka and Lamb 2005, 132, 248)

Plants exhibit impressive selection mechanisms, not only to avoid self-pollination, but after fertilization, aborting some of their fruits to strengthen others that are more promising. Robert Bertin discovered a plant *Campsis radicans* that is pollinated by one donor only as it removes pollen from other flowers. He notes that the seeds that grow from selected pollen produce hardier plants and that plants exercise selection in looking for cross fertilization with farther or more remote samples. (Trivers 1985, 344–45)

Here's a successful courtship manual for male birds:

1. Choose a point, because when the mating is successful, one tends to return to the same point.
2. Prowl females in circles, at a distance, to avoid them leaving your territory.
3. Sometimes try to reject the female during courtship, turn your back against her insinuating that there is a time limit to respond.
4. Be timely to approach the female, surround her from a distance and go closing the circle without hurrying. This has been proven in the case of the black grouse (*Tetrao tetrix*). (Trivers 1985, 349–51)

These observations on birds seem equally useful to teenagers and adults in the human species. An overeager male indicates that he has been lonely for a long time, which detracts a candidate because the more confident implies that he is more desired and desirable.

At the organismic level sensory terminals and a nervous system are required to perform the third selection mechanism that depends precisely on the perception of phenotypic characteristics. This selection appropriates behavioral habits in hunting and fishing whether by frequent stimulus—as frequency of finding sticks near termite nests as explained (Galef 1992) or by the dominion of individuals in certain species whose behavior is mimicked (the case of potato washing by Imo the chimpanzee).

Cultural selection, in addition to including previous practices of imitation and learning, covers selection mechanisms from the *secundus* such as faithful reproduction or copying objects in permanent formats, maintenance of codes and texts to facilitate their re-appropriation and interpretation, accumulation and collection of resources that are considered valuable.

Roger Bartra (2007) proposes understanding culture as an exo-brain, and Dawkins's genetic determinism sees it as an extended phenotype that contributes to the ability of carriers to replicate their DNA. Just as the genotype uses bodies for replication machines in a serial genetic industry, the meme uses brains as carriers and vehicles. Memetics' main advocate is Daniel Dennett, followed by Susan Blackmore (2000) who tried to develop Dawkins's (1999) idea on the basis of imitation. However memetics does not explain how such a mechanism evolved (except that it is somewhat of a secretion of the human brain), how cultural variation is generated, what the nature of cultural change is, which memes are selected and on what criteria, or how these memes jump. (Blackmore 2000) There are no hints at how memes are transmitted, stored, or interpreted, and there does not seem to be room for animal culture because it presupposes that the meme is always human (although Reader and Laland do suggest the possibility of animal memetics [1999]). Books are meme replicating devices, but how they coagulate memes into writing and then re-coagulate them into thoughts, we do not understand. Despite the vagueness of how it is defined and activated, and how it jumps to another brain, the meme of the meme seems highly contagious. Until today, such a proposal has been best developed by Robert Aunger (2002), who, instead of using minimal units, proposes memetic complexes or memeplexes by electric charge mechanisms through neuronal networks, or what he calls "neuromemetics." I remain expectant to see future results on this theme.

Transmission Devices

In genetic transmission, for Pert: "Amino acids are letters. Peptides, including polypeptides and proteins, are the words made from these letters. And they all come together to make up a language that composes and directs every cell, organ and system in your body." (Pert 1999, 65)

Several genetic transmission devices have been found such as miRNA, tRNA, rRNA, siRNA, mRNA, and other RNA associated to protein Piwi. We already mentioned that at the next level, the transmission devices are epigenetic marks that activate or block chromatin information in stem cells by methylation. These marks refer to hereditary components of chromosomes regulating the activity in cell lines transmitted to daughter cells.

Among the consequences of various forms of transmission are recognizing that:

> Variations in socially transmitted behaviours affect evolution in two ways: first, the variant behaviours are an additional source of raw material for selection; second, social behaviour forms part of the selective regime in which individuals live, learn and reproduce. Therefore, habits and traditions are not merely the products of evolution—they are also one of its major constructing agents. (Avital and Jablonka 2000, x)

> As we see it, any evolutionary interpretation of social behaviour requires a consideration of both genetic and "cultural" factors. Variations in socially transmitted behaviours affect evolution in two ways: first, the variant behaviours are an additional source of raw material for selection; second, social behaviour forms part of the selective regime in which individuals live, learn and reproduce. Therefore, habits and traditions are not merely the products of evolution—they are also one of its major constructing agents. (Avital and Jablonka 2000, ix–x)

In the transmission behavior among various species such devices are the parents themselves and other contemporaries of the recipient. Also, the group itself can function as a transmission device reproducing modes of conduct by the frequency of the stimulus. At the cultural level transmission devices are teachers, parents, priests, speakers, leaders, managers, publicists, idols, models, celebrities, and so on. Their strength depends on the closeness, hierarchy, and frequency of contact as well as on the degree of subject's dependence to these agents of cultural transmission. We should include as a device the format or physical transmission vehicle such as media in all its visual, auditory, textual, gestural possibilities: television, film, language, ceremonies, laws, and customs that place recipients as captive audience for transmission.

Transmission Mechanisms

The organization of DNA molecules is transported by the RNA to the amino acids in the environment for the formation of new bases. This genetic transmission mechanism requires a chained code of nucleotide molecules that replicates or copies an equal structure for the formation of peptides and proteins. As Pollack (1994) notes, copying occurs by means of contact of the source of the transmitter to the receiver. At another level, the epigenetic transmission to the daughter stem cell operates by the silencing of certain genes where methyl groups (CH3) adhere to the bases of DNA and chromatin parts affecting gene activity of each cell by methylation according to location. How they choose which genes to methylate is not yet understood.

What we do understand better is the mechanism of transmission of emotions and action in empathy and imitation since the discovery of mirror neurons by Gallese and Rizzolati.[7] It consists, as shown by fMRI studies, of the activation of neurons in observing an action in the same exact area that would have been activated by performing such action. We identify with what we observe and virtually perform in part the action at a neuronal level. The mirror neuron system could offer a basis for what Krebs and Dawkins defined as "mind reading" namely to read others' intentions by their nonverbal expression. From this perspective, when we hear a symphony we not only perceive the sound but virtually almost experience playing it by vicarious identification with the musicians. The artistic experience may well consist of feeling vicariously the skill and emotional intensity invested in painting by Van Gogh, singing like Diego El Cigala, dancing deftly like Nijinsky, or playing Bach on the piano through listening to Glenn Gould. We obviously experience art in our own emotional and corporeal terms, placing ourselves in the artists' shoes by in-corporating (em-bodying) an artist's sensitivity, speed, elegance, fervor.

Luca Cavalli-Sforza and Marcus Feldman produced mathematical models to describe and observe processes of transmission of cultural practices over time and analyze them in quantitative terms according to statistics. According to the authors there are three types of cultural transmission: vertical (parents to children), horizontal (colleagues and friends), and oblique (priests, models, or other authority figures). (Cavalli-Sforza and Feldman 1981) The oblique transmission of cultural accumulation allows us to learn from Socrates, enjoy Boccaccio's tales, and admire horses at the Chauvet cave from 30,000 years ago. The current media transmission triggered horizontal inheritance to previously unimaginable exponential levels, networking us with our contemporaries in spatial and cultural distances by internet and twitter.

Language is crucial as a mechanism of cultural transmission. There are other factors such as exporting behaviors and know-how like the use of tools that can radically alter the way of life of a group without the need to communicate linguistically. A case in point is the use of the stirrup, the compass, and gunpowder imported from China to Europe that revolutionized economy and military tactics. (Needham, Robinson, and Huang 2004) Kroeber is a pioneer in applying an evolutionary perspective to culture by highlighting the diffusion of systems or cultural traits (ideas or stimuli) such as the European reinvention of Chinese porcelain in the eighteenth century to save import costs. (Kroeber 1917; 1940) In each of these cases, where diffusion operates in space as transmission in time, the context has a role not only in the selection of the most suitable trait for each environment, but the frequency and disposition to acquire certain available models and not others.

In analogy to the genetic code that regulates the nucleotide sequence in replication, cultural forms regulate social activities and their reproduction by protocol codes in diplomacy, etiquette, as well as in ostensive expenditure systems like mayordomía, potlatches, or wedding rituals. Distinctive features in the transmission mechanisms of *tertius* are on one hand its deliberate, not only random, heritability of individual findings, namely accumulation of transformations or inventions, and on the other, high speed transmission that does not require the passage from one generation to another.

Selection Bias and Natural, Behavioral, and Cultural Transmission

According to Darwin we can consider two types of selection: natural and preferential (sexual). In the first case selection is performed not by an agent but by the context of traits that better contribute to the survival and reproduction of the bearer. In the other there is an additional factor that depends not in the organism itself and its suitability to the circumstances but on a re-actor, namely some other creature that values, prefers, admires, imitates, and chooses a trait or variation in sexual selection. Whereas natural selection is determined by the perceptor-effector semiosis in decoding circumstances and possibilities when an individual reacts accordingly, in sexual selection semiosis introduces actor–re-actor relation in a social circumstance. We have here an intention by the actor to impact the re-actor for achieving a particular goal. The presence of each re-actor significantly alters the configuration of the message seeking to impress it. These valuations and preferences depend on adaptations and can be automatic or deliberately quality selective as in the case of the spider *Paratrechalea ornata* where the nuptial gift of this species is composed of a prey item wrapped in silk having visual and/or chemical cues involved in inducing female behavior. (Brum, Costa-Schmidt, and Araujo 2011) They found out that males can build low-quality nuptial gifts consisting of already digested prey or inedible items, like a seed wrapped in silk where "digested prey wrapped in silk is readily accepted by females, but digested prey without silk is never grasped by females." Such discrimination, they argue, is mainly chemical rather than visual.

The condition for the occurrence of preferential selection is the evolution of a well developed nervous system that is able to discriminate between minor differences by discernment between one individual and another. There are inherited and acquired preferences in selection.[8]

There are three types of biased transmission: (Boyd and Richerson 1988, 10)

1. Direct bias as in the case of children who receive cultural information from the family and are predisposed to acquire it for their location, as in diet preferences.
2. Indirect bias where the individual chooses for reasons unrelated to cultural traits themselves but on terms of success, status, or wealth of individuals taken as models and imitates their dress, speech patterns, and other features which at first glance seem to have to do with obtaining prestige and power.
3. Frequency dependent bias refers to cases where the individual acquires cultural variants and traits that are the most common in the environment. This bias or tendency often manifests itself in forms such as consensus, conformity, political correctness, or bandwagon effect (following the general opinion). Its inverse mode, call it rarity bias, chosen precisely for being unusual, is what the elite follows to distinguish itself through fashion, avantguard arts, the eccentricity of style, or an unusual skill to have better employment opportunities. (Richerson and Boyd 2005, 121) In fact, the Drosophila fly was found to have a preference for less frequent or minority males in the area searching for greater genetic variability. (Trivers 1985, 337–38)

For Richerson and Boyd (2005, 69) biases can occur in two ways, innate preferences and cultural preferences acquired by social learning. Conformist bias is useful when the environment changes very little and there is a very small risk of mistakes, which then tends to reduce the amount of variation within groups while it increases between groups. (Richerson and Boyd 2005, 162) It also requires less effort. In addition to numerous experiments that prove this tendency to conform to views according to the local moral, this trend is more prominent in political positions. (Richerson and Boyd 2005, 122) The ideological packages that people adhere to with surprising docility, even (or particularly) groups considered "critical" as so-called "intellectuals," offer an all inclusive set to vote for a candidate of dubious reputation as the candidate of "left" or "right." Such support implies following consensus among friends and peers, favoring any initiative that comes from the biased side and repudiating the opposite. Conformist bias is always ultraconservative whether leftist or rightist, since it is subjected to the group's inertia as a result of the adaptive advantage of being part of such group. (Richerson and Boyd 2005, 163)

Boyd and Richerson's categories have been further adjusted by Henrich and McElreath when they distinguish "content bias" that chooses cultural traits by their characteristics themselves (i.e., direct bias) and "contextual bias" of two types, the frequency dependent bias and indirect bias based on models arising from four biases: (1) prestige, (2) skills, (3) success, and (4) similarity. (Henrich and McElreath 2004)

Evolutionary researchers and psychologists have characterized several different types of biases through a spectrum from individual to social scales, as the status quo bias where in a disproportionate manner people tend to make decisions just following the status quo. (Fernandez and Rodrik 1991) Common are also least resistance biases, conformism or herd behavior, and occupational biases. The occupational preference may be affected by hormone levels, as in the study by Dabbs, de la Rue, and Williams (1990) which compares levels of testosterone among actors, athletes, and ministers (whose levels are always lower than the former). Climate seems also to impact preferences, as noted in the study by Barber (2001) and the relative proportion of single men that drives them to emphasize their masculine traits such as beards, sideburns, and mustaches in the difficulty of obtaining a mate.

These biases are always due to practical reasons as to avoid losing the support of the group and the convenience of following what is habitual to overcome firsthand mistakes. Researchers in neuroscience and experimental psychology found specific trends in human perception (perceptual biases) regarding various aspects perceived in objects such as faces' symmetry, age, gender, hormonal saturation, all of which apparently depend on specialized brain modules. One more category needs to be added to the taxonomy of biases, namely the aesthetic bias which we will examine in the last section of this chapter.

Variations or Generation of Changes

For selection to operate there must be variation; and for evolution to operate there must be selection and transmission of change. Variation in the *secundus* is almost null in asexual organisms as they repeat the same elements by mitosis; by contrast, in meiosis there is a recombination and shuffling of elements that configure a greater variety of organisms. The variation is even greater in the *tertius*, for not only elements are recombined but reinterpreted, sometimes with totally unpredictable results.

Most of the variation in the *tertius* arises from competition for social and economic survival, especially in densely populated areas, trading goods with both material and semiotic use and exchange value. Due to pressures and drives in bio-cultural evolution, a number of these goods signal fitness in the presentation of identities and competition for social status (brands of clothing, cars or phones, pets, tastes in entertainment). Three types of cultural capital are operating: monetary capital, social or contacts capital, and educational or professional capital that can influence and transform one another and affect the frequency and quality of a trait.

A neutral biological mutation can keep replicating through several generations, but a harmful mutation can affect the chances of survival of the host organism in a certain context or improve them in another and favor replication. By contrast, in cultural heredity, harmful cultural phenomena such as xenophobia, misogyny, racism, child abuse, and the demonization of ethnic, national, or religious groups (such as romaphobia, israelophobia, homophobia, and alterophobia) can remain endemic or recessive over a period of time and all of a sudden burst and increase exponentially by contagion or bandwagon effect. We have witnessed dramatic cultural changes of epidemic cultural transmission in the twentieth century during its massive conversion to Marxism in Eastern Europe and China, the Nazification of Germany for a lethal decade and a half, the Westernization of Japan after the Second World War, the orientation among baby boomers and hippies towards cultural traditions of India, Japan, China, and Tibet during the sixties and seventies, the present muslimization of Europe, the cybernetization of the world during the nineties as well as the recent massive facebookization.

The difference in evolutionary speed among the three areas is abysmal: the vertiginous exponential change in the *tertius*, in contrast to a slow evolution of species in the *secundus* and then the relatively static quasi-Aristotelian fixed glass of the *primus* universe after big bang's radical changes in fractions of a second.

Evolutionary Forces

"By 'forces' we mean causes of cultural change, analogs of natural selection, mutation, drift and so forth in the genetic system of inheritance" Boyd and Richerson point out. They highlight five forces in the mechanics of cultural transmission:

1. *Random variation* that occurs due to transmission mistakes, such as mutations in genetic variation: "Our intuition is that the rate of cultural mutation is much higher than the rate of genetic mutation. If this is so,

accidental variation may play a somewhat different role in cultural evolution than it does in genetic evolution."

2. *An analog of genetic drift* "If the population of models active in cultural transmission is small, chance variations in which cultural variants are observed and remembered may cause substantial changes in frequency form time to time."
3. *The force of guided variation*, "the effect of guided variation force on evolution depends on the existence of some adaptive standard such as taste or a sensation of pleasure and pain."
4. The *biased transmission* such as *content* bias, *frequency* and *model* bias.
5. *Natural selection in culture*. "Natural selection acting on cultural variation can cause the evolution of different behaviors from those one would expect as a result of selection acting on genetic variation when the pattern of cultural transmission is different from the structure of genetic transmission."

(Boyd and Richerson 1985, 2, 9–11)

Just as there is a particular genetic pool in *secundus* for an isolated population or ethnicity, there are local pools in the *tertius* where certain values and role models are more abundant, fertile, or contagious. If we extend the metaphor, some cultural traits can be recessive and others dominant.

Richerson and Boyd (2005, 51) alert "we urge great care with loose analogies to mutation and selection because several distinct processes rooted in human decision making lead to the accumulation of beneficial cultural variations, each with a distinctive twist of its own and none exactly like natural selection."

Let us consider a sixth evolutionary force: chance or catastrophes. The K / T mass extinction of the Cretaceous-Tertiary or Chicxulub meteorite that hit Yucatan 65 million years ago with the extinction of reptiles and several plant species (which birds and mammals survived), enabling the emergence and survival of other species dramatically illustrates this sixth evolutionary accidental force.

Crosses of Co-Evolution

Boyd and Richerson (1985, 2) seek to unite cultural transmission models to genetic evolution models "to determine the circumstances under which natural selection might favor the modes of cultural transmission observed among contemporary humans. These models show that the same modes of cultural transmission that led to novel evolutionary results can arise via conventional neo-Darwinian processes." They assert that cultural factors may determine the selection of mental and physical aspects—that is, when there is a direct interaction circuit from *secundus* and *tertius*.

Durham's study on lactose tolerance proves this feedback between culture and genetics and their co-evolution, because it shows that humans evolved to metabolize lactose only in early ages but stop metabolizing it in maturity, so lactose intolerance in adulthood is an adaptation. However, in areas of the northern hemisphere there is the highest frequency of individuals with genetic tolerance to lactose because they are dedicated to farming and strong dairy consumption. This can be explained by the fact that in these areas sun exposure is less than required to fix calcium, which is compensated by selecting genes that favor the allele to digest lactose and co-evolve with animal husbandry practices. (Boyd and Richerson 1985, 159, 162)

Durham's work on gene-culture co-evolution shows that while there is a difference between the two orbs, they are interwoven through meaning and interaction webs. Therefore we should design experiments that show the dynamics and conformation of these crossings of natural selection in culture, as well as of cultural selection in nature like cases of pollution, cloning, extinction, and recovery of extinct species by DNA. Another aspect of crossing is sexual selection of certain phenotypes and marginalization of others according to fashions in preferences, as in thin or robust, round or angular features in female anatomy.

As noted also by Jablonka and Lamb, there is a flow in both directions between different hereditary systems since "The reciprocal influences of the two heredity systems can be seen most easily in human populations, but they are not limited to humans or even to social animals. Whenever the activities of one generation shape the conditions of life for the next, there will be feedback between the inherited genes and the inherited niche." (Jablonka and Lamb 2005, 293.) The influences of *orbis tertius* upon *secundus* are multiple and increasingly extensive: increased longevity by the progress of medicine, obesity and anorexia by the glamorization of skinny models and sizes only for the undersized, increase in cancer incidence by teratogenic food production, hypertension, stress and anxiety due to overcrowding social conditions of contemporary urban life, overexploitation of sexual drive by advertising for consumption, and pollution of natural resources by industry and subsequent genetic mutations.

In this chapter I have just managed to sketch a subject that obviously requires further investigation to distinguish and integrate the components and dynamics of evolution or, in some cases, not only the evolution but the involution of the *Orbis tertius*.

THE OTHER GARDEN OF FORKING PATHS

Just like nature in the *orbis secundus* is a garden of forking paths through species' clades or lineages derived from common ancestors, so is culture in the *orbis tertius* whose paths branch off through matrices and submatrices that germinate from a seminal protomatrix. Thus, by focusing on the *tertius* in more detail, one can witness that such vast cultural production actually ramified from a relatively simple and common origin: the maternal link to her offspring, becoming family, clan, tribe, nation. The most conspicuous areas of the *orbis tertius* are the cultural matrices, equivalent to the monera, protista, fungi, plantae, and animalia in the *secundus*.

Already from the *orbis primus* Lee Smolin (1996, 298) points out that in every black hole there is a region where one can find the birth of a new universe retaining qualities of a previous "parent universe." We know that in the *secundus* every creature is engendered from a mother cell or a gamete whose qualities it retains. In the *tertius* matrices also germinate from earlier ones that inherit some components and further diversify to respond to emerging situations and demands. Hence this is the case of symbols that evolve not by landing from above upon society as cherubs to announce the immaculate conception and birth of new concepts, but by growing from below and reflecting human needs and vulnerabilities in confronting disease, mortality, coexistence, and catastrophes. Social symbols thus strengthen cohesion and action for the protection and reproduction of the individuals and the group. Where a symbol is brewing, it magnetizes the whole area which is revitalized attracting other cultural webs that latch into it like axons looking for dendrites. With the passage of time these areas accumulate symbolic density by repeated flow and gradually evolve to increase their material, energetic, affective, laboral, and political charges.[9]

Matricial Seeds and the Genealogy of Culture

Lynn Margulis discovered that "different bacteria form consortia that, under ecological pressures, associate and undergo metabolic and genetic change such that their tightly integrated communities result in individuality at a more complex level of organization. The case in point is the origin of nucleated (protoctist, animal, fungal, and plant) cells from bacteria." (Margulis 1996, 136) At the cultural level, matrices similarly appear to be a result of the association of individuals under ecological pressures that generate common practices resembling the genesis of eukaryotes. Individuals join to form a matrix, whereas matrices separate and differentiate themselves from the mother matrix through a process of specialization, as do offspring from parents.

Each cultural matrix takes root and grows from the symbol that gives it coherence and from which it branches out by means of differentiated verbal, acoustic, somatic, and scopic (visual) practices. Each matrix projects configuration aspects of performance onto others through patterns that organize their practices in shared reticular realities within their cultural environment (as flowers when exchanging pollen and mimicking the colors of neighbor flowers chromatically tuning with them in the *secundus*). Such reticular structure implies that there always are submatrices within each matrix, which in turn generate others as fractals at various scales. The evolution of social matrices propagates semio-aesthetic cultural variants to demarcate differences and generate rules that favor certain social dispositions and punish others, enhancing particular organization patterns of preferences.

The British embryologist Brian Goodwin (1996) suggests the idea of morphogenetic fields for the development of forms as complex as the eye that cannot be explained by natural selection of random variation. This idea of formal conditioning fields could be related to the concept of convergent evolution that Salthe illustrates with the parallelism of very different creatures like seahorses and chameleons having the same style to hunt called "bushwhacking" while hidden in vegetation, and suddenly jumping. The parallel extends further, as both have strange colors, prehensile tails, and eyes that move independently. (Salthe 2008, 140–41) We can also find parallel independent convergences not only among animals, but in cultural matrices, as handicrafts' patterns common across various locations, particularly in ceramics and embroidery that emerged separately by requirements of the materials used.

Applying a biological model to the *orbis tertius* we can consider it as evolving by pregnancy (seminal ideas of a technical or scientific invention), growth (development), maturation (crystallization), reproduction (diffusion), and submatricial branching (creating further variations). A critical population mass is necessary in a cultural matrix to provide stability to its practices and beliefs and guarantee their reproduction. The matrix as a symbolic field, requires particular agents close to its center, who understand the value of the unifying symbol, for its collective autopoiesis and reproduction. What Gramcsi called "organic intellectuals" are here pertinent to understand the reproduction and legitimation of a matrix within the social network (such as priests in the religious matrix, shamans in the tribal, actors in the artistic, athletes in the sport matrix, researchers in the scientific matrix, teachers in the educational, and bureaucrats in the state matrix).

Matrices can be heterarchical developing horizontally as rhizomes (in Deleuze and Guattari's (1966) use of the term) or hierarchical and vertical as arboreal configurations, but their autopoiesis always requires feedback loops from the environment and therefore porous boundaries or membranes to separate the matricial "us" from the "them." One can clearly identify each matrix by the specificity of its production in each register and the limits it established (e.g., documents as property certificates or physical barriers such as national boundaries, dividing walls in churches, prisons, houses, cemeteries, offices, schools). Some membranes are lexical as family surnames, discursive conventions, or legal jargon that exclude those not belonging to it;

others are set in time as dates for conferences, weddings, and carnivals in addition to visual or scopic membranes to distinguish and separate the members, for example by their dress (caftan, burqa, toga, uniform, or Armani suit).

Unlike the Marxist concept of "superstructure" as a scaffold built over economic life, culture does not reflect the structure since it is its very fabric in daily activities that enable the production of goods and techniques, practices and values for social and individual survival. Economy is metabolism and work, both in the *orbis tertius* and in the *secundus*, flooding everything as one of the key dimensions of life. Even in the *primus*, the law of entropy is an economic law involving effort and waste by measuring how much energy is available to exert work in a closed system.

As metabolism in a cell, the matricial network in culture maintains and absorbs energy and material resources by its nodes through the flow of work, discursivity, and affectivity. A matrix requires memory kept in physical compilations akin to the genetic and immunological history of an organism. Schools keep educational records, families keep albums and anecdotes, lovers their love letters, scientists record conferences in proceedings and papers in journals, the military keep records of incursions and deployments, the state keeps a record of taxes and duties, and each nation keeps its oral and written history.

The socialization of individuals depends thoroughly upon matrices. Consequently, there can be no identity which does not participate in a cultural matrix from which it organizes its meaning and context, its signs and symbols.[10] There is no culture lacking of matrices, even if concentrated in a single one from which the rest spring forth. In the Pleistocene the matrix was the tribe upon which the members' political, magical, ritual, semiotic, aesthetic, and sexual practices as well as economic livelihood and mental configuration depended. At present, totalitarian regimes like Iran, Nigeria, and Saudi Arabia compress the large and diverse social matricial fabric into a single monolithic block that controls every single aspect of people's life, imposing its rules on the whole matricial network: family, school, entertainment, science, politics, language, and so on. This matricial clustering is a maladaptive involution that is lethal, as it destroys distinctions and practices that have taken millennia of evolution and human wisdom to evolve.

Since it is we as individuals who are the agents and builders of our socially constructed matrices, we have direct access to them and can understand and transform them taking into account differences of parallax that bias individual and local particularities. I see no contradiction between social constructionism and evolutionism, contrary to Tooby and Cosmides (1992, 81) who neglect this crucial factor that literally builds culture as we build a book or a house: piece by piece, with concrete materials, techniques, and ideas. Therefore, no matricial networks can be seen as systems independent to the observer since each matrix generates a perspective for the signification of events, framing it according to the subject's contextual position. From each matrix, a particular mode of seeing and feeling the world is bred as the tissue with which we articulate a shared reality on the basis of agreements, training, and practices that configure every individual as a social being sharing a collective identity.

Matricial Branching, Sprouts, and Grafts

To study each matrix separately it is necessary first to observe its autopoietic processes and feedback loops such as rituals and customs that recharge its symbolic power through accumulated time becoming inertial. The preservation and transmission of anecdotes and the safeguarding of memory in each matrix, the regulation of activities in time, the production and use of material and mental resources are analogous in the *tertius* to fermentation, photosynthesis, respiration, and nitrogen fixation in a plant.

In *secundus* as in *tertius* and *primus*, space is never flat given the fact that places are unevenly distributed through a network of hierarchical configurations similar to mountains and valleys, forests, watersheds and craters. Matrices endow and consecrate meaning to places by contiguities and contrasts of identities, ranks and social roles in usually concentric territories. In the animal kingdom, the bravest individuals often monopolize the center because more links converge there than the periphery.[11] Birds increase their aggressiveness and displays right in the middle of their territory, as the male sparrow starts the day with a song at the center of a *lek* to assert himself.[12] The grouse moves towards the center of the *lek* as he gains higher status, because most females are located there. Hierarchies take place among females, for example between white Leghorn hens whose resemblance to the former leader, fighting skill, aggressiveness, and even luck determine their status.[13] Among the macaques, bonobos, and papios, the status is passed on to the offspring, but in other cases is obtained by body size or rudeness. Since status is the guarantee of access to the necessities of life such as shelter, food, leisure, and sexual partners, this drive for status passes from the *secundus* to *tertius* through matrices that administer recognition, protection, or exclusion.

Not all primates are social species: the orangutan, which has a large body, lives alone, while the bonobo having a smaller body lives in matriarchal societies or tribes of nearly a hundred. Anatomical weakness of Homo sapiens explains, for Darwin, the selection pressures that forced us to assemble in social groups to survive.

> We should, however, bear in mind that an animal possessing great size, strength, and ferocity, and which, like the gorilla, could defend itself from all enemies, would not perhaps have become social: and this would most effectually have checked the acquirement of the higher mental qualities, such as

> sympathy and the love of his fellows. Hence, it might have been an immense advantage to man to have sprung from some comparatively weak creature. (Darwin 1882, 64)

Elephants, however, are social beings, despite their big size, so volume alone cannot be the whole criteria for the socialization of particular species. According to Trivers (1985), our culture derives from computing and remembering links and social identities to fight cheaters and freeloaders. Reciprocity is a social requirement expressed in mutual grooming and submission to the alpha male among chimpanzees in exchange for keeping order and avoiding conflicts within the group. This sense of animal reciprocity is so deeply rooted in the *secundus* that it extends to the *tertius* as expressed by the ritual of sacrifices to the gods, ostentatious expenditure, and sacrificing a portion of the harvest in return for new crops (as we will see in chapter 5).

Matricial Gamete: Proto Matrix of Gender

> *So God created man in His own image, in the image of God He created him; male and female He created them.*
>
> —Genesis 1:27

In the branching process of cultural matrices, the first distinction is the difference between male and female. Ever since the *secundus*, male and female in several animal species are distinguished by their distinctive anatomy, reproductive function, behavior, and decoration. It is easy to recognize the male in multiple species by their ornaments, larger size, and build to compete against rivals. Among chimpanzees, however, the difference is expressed less in muscular arrangement, since females mate with several males during their fertile period and competition takes place inside their womb.

Gender dissimilarity is not only anatomical but psychological as well, according to Kawai's (1965) experiments on psychomotor skills. Matsuzawa and Yamakoshi (1996) indicate that females chimpanzees tend to be more witty and skillful than males and more dynamic in migrating from one community to another. Customs to obtain resources and division of labor in hunting and gathering derive from these differences, as the case among papios where the father takes the role of silently disciplining the offspring. (Wilson 1980, 146; Trivers 1985, 2)

It is significant that human beings are the only species where rape or forced copulation takes place, quite rare in nature as noted by Trivers. He argues that females in animal species may resist unwanted copulation with males by either as seals turning around, or balancing the body, flight, changing territory and many more resources. (Trivers 1985, 341) Statistics show that males of our species carry more infections than females and die more frequently in homicides, suicides, and accidents (they are more likely to drive recklessly as shown by Daly and Wilson's study in Canada in 1976 where the risk is significantly increased from age fourteen to thirty-eight). (Trivers 1985, 303; Daly and Wilson 1983) In fact, it is questionable to what extent the culture of masculinity is responsible for the disproportionate violence by encouraging male gender stereotypes, as not everything can be explained only by the generation of testosterone.

Figures from the British Police Crime Survey in 2009–2010 show that men were responsible in 91 percent of violent incidents in England and Wales and vary by type of incident: 81 percent for domestic violence, 86 percent for assault, 94 percent of injuries, 96 percent for mugging, 98 percent for robbery. In the Ministry of Justice for 2009, figures show that men are responsible for 98 percent, 92 percent, and 89 percent of sexual offenses, drug offenses, and criminal damage respectively. Of sex offenders, 99 percent are men. The highest percentages of female offenses concern fraud, forgery (30 percent), theft, and possession of stolen goods (21 percent women).[14] According to crime statistics in 1959 social scientist Barbara Wootton observes that "if men behaved like women, the courts would be idle and the prisons empty."[15]

The morphology in signaling and gender behavior extends and projects from *secundus* to the *tertius* with significant variations across cultures. Despite gender being a proto-matrix, conventions in the presentation of identities are already clearly displayed by aesthetic strategies in the four registers. In the scopic or visual register there is no culture or tradition that does not prescribe the use of differenced clothing, jewelry, decor, tools according to gender. In the acoustic register, voice modulations are established such as exaggerating the high pitch in females' voices as in the use of a different gendered vocabulary in Japan. Rumi Washi (2004) describes how in the Meiji period from 1890 to 1912, a special effort was deployed to "feminize women" by instituting a "female speech." The matrix of the state through the Ministry of Education, along with the National Association of Language (Kokugo Shingi-kai) imposed in 1941 this modality during the Showa Era (1926–1989) of specific vocabulary and grammar. In addition, a manual of etiquette was elaborated as part of an ideological effort during the Second World War with the alibi of supposedly strengthening national values during wartime. The aesthetic presentation of female identity in Japanese women by the kimono that significantly limits the movement of the body and gestures is a paradigmatic case of this intersection of two orbs. This case recalls the former custom in China of tying girls' feet to prevent normal growth, or of today's burqas limiting Muslim women's natural perception and normal peripheral vision, in addition to the horrific custom of clitorideal ablation.

This gender proto-matrix has been progressively controlled and dominated not only by the state and family matrix, but also by the religious, commercial, medical, and juridical matrices and very often implemented to oppress women in their laboral, political, and semiotic power by controlling their fertility, mental and sexual potential. The feminist debate on the so-called nature vs. nurture dilemma requires taking into account also its liminal locations within both *secundus* and *tertius*, exactly like a bulb nascent between earth and heaven, and between nature and culture. Gender is rooted in the evolution of a morphological difference from *secundus* not only genital and hormonal but neuronal and psychological. In the *tertius* that root extends to the devices used for marking gender identities in each culture to preserve endogamy to the degree of imposing genital, mental, perceptual, and facial mutilation upon half its population. Such impositions can lead to social, aesthetic and semiotic consequences such as establishing the subordinate role of the women, imposing rules on their behavior and limits to their growth. Matricial practices are initially focused on survival but may become maladaptive.

Matricial Stem: The Family

The family is the unit of gestation, production, feeding, breeding, socialization, tradition, acculturation, and memory protection of the human species. Being so close to the biological origin, the paths of *secundus* and *tertius* merge through the family into genetic and behavioral transmission, because biological parents are usually also the first cultural parents. The family matrix is a cultural evolution from gender protomatrix rooted in anatomical difference by establishing specific roles for the sustenance of the offspring.

From the smallest unit of the couple to the family, the clan, the tribe, the chiefdom to the nation, every matrix marks boundaries and frontiers between the "us" and the "them" that start from the recognition of one's offspring. While distinction in the gender protomatrix is individual, the family matrix triggers emerging and expanding groups with increasingly complex collective differentiations.

The process from *significans* into *cogitans* can be explained in part as a result of the need to avoid incest by identifying individuals and keeping record of acts of rivalry and loyalty within and between groups. The incest taboo is not exclusively cultural as can be traced already among bonobos, and even plants that refuse to be fertilized with their own pollen, as we mentioned in the previous section dedicated to selection mechanisms. (Trivers 1985, 344–45)

Genetic memory inscribed upon our cells in *secundus* extends here to memetic memory stored in *tertius* in stelae, scrolls, parchments, libraries, and hard discs. The juridical matrix establishes rights, obligations, property, and inheritance according to the genetic template. The religious matrix consecrates or prohibits the mating of its members according to its rules and registers every member from birth to death according to family's affiliation. In the Bible we have a unique document that accounts for the evolution of a family matrix to tribal, national, religious, and state confederation of twelve tribes in the kingdom of Israel with its capital in Jerusalem three millennia ago.

Every culture records mythical founding parents (Abraham and Sara for the Hebrew nation, Romulus and Remus for Romans, Brutus of Troy for the British, Huitzilopochtli and Coatlicue for Aztecs, Jefferson, Washington, Franklin et al. as the "Founding Fathers" of the United States, William the Silent for the Netherlands) and keeps family relationship schemes that define identities within and among clan and tribe members. The constant in all anthropological and ethnological studies are these kinship records that mark forms of exchange and hierarchy, descent and division of labor with obligations, rights, and exigency of reciprocity. As cellulose is the physical structure in plants, kinship is the basic matricial structure for growth in the *tertius*. Since the family is rooted in common genes, the link between the two orbs is salient. As a tendril of *Clematis flammula* plants detects where to attach and grow, the pater familias has the role to discern in which direction the family should move and settle.

It is no coincidence that all drug cartels in Mexico—like the Beltran Leyva and Arellano Felix family—directly emerge from the family matrix seeking to impose its own law and to legitimize illegality. These *narco*-families operate their own "legal code" of bloody revenge, and employ aesthetic strategies such as the musical genre of *narco-corridos* of songs praising their adventures. Monumental narco-architecture is also worth mentioning as it ostentatiously mimics fairy tales or imperial and historical styles like the ex-police chief Durazo's Acropolis or the "Palace of the Arabian Nights" in Guamuchilito owned by Amado Carrillo Fuentes, the so-called Señor de los cielos. This man hired his *narco*-medicine to change his *narco*-face and, after having murdered hundreds of policemen and being hunted by national federal police and international Interpol, ironically died from plastic surgery in 1997.[16] There are also narco-religions, such as "Jesus Malverde" a local thief made holy, or the "Familia Michoacana" cult (with more than five thousand members and the cause of more than twelve thousand deaths), whose leader Nazario Moreno Gonzalez nicknamed "El más loco" (the craziest one) used to preach Bible in hand a bizarre New Age religion based on the book *Wild at Heart* by John Eldredge. These organizations receive popular support because they perform functions that the state is unable to perform, such as funding schools and churches, loaning to farmers, and financing drainage projects. These drug cartels illustrate the process of emergence of social matrices.

The ideological metaphor of the "big family" so commonly used in big business corporations, religions, and political parties takes this paradigmatic projection to manipulate and obtain cohesion. This figure is also used to propagate nationalist ideologies and sports affiliations since genetic affinity is the most effective structure to mobilize emotions. From the Pleistocene, affections such as loyalty, reciprocity, shame, revenge, and honor are inscribed in our genetic, memetic, and emotional evolution, hence their enormous impact and disturbing force.

ARTS, ARTIFACTS, ARTIFICES

The first duty in life is to be as artificial as possible.
—Wilde

With the radical climate changes that occurred during the Pleistocene two possibilities were open: either hominids lacking thick fur and fat would be extinct by cold during the ice ages, or hominids with thick fur and fat would be extinct by heat during periods of high temperature and drought in the savannah and desert. We are naked apes, as Desmond Morris calls us, because we made the best choice of blocking adaptations of fat and skin fur that protected us from the cold. We did it with cunning in a highly variable and extreme climate context by inventing the artifice of covering ourselves with other animals' fur or fat and defending ourselves to survive the cold and also surviving the heat just by taking furs off and on whenever necessary. Culture is thus the rule of artifice.

There are at least four aspects that stand out when we try to define culture: it is artificial, hereditary, cumulative, and collective. Beavers pass on to their descendants the artifice of their branch dams, and rats pass on nests to their families. The bumblebee queen can use an empty mice nest for building a dome over it to form a honeycomb in which she deposits a cluster of buds for future workers. Also, crocodiles, caimans, and alligators build nests with leaves, sticks, and branches to lay their eggs on water, and even use their decomposition to increase the temperature. (Wilson 1980, 201–2, 212) Mirmelontid lion ants dig funnel-shaped wells in the sand to catch prey, and caddis flies' larvae make silk cases where they stick sand or plant debris. Given the fact that we have a clear case of liminality or fuzzy border between *secundus* and *tertius*, we must consider the possibility of proto-cultures in plants and animals. Salthe notes that:

> And there are trails in the mountains that have been used by deer for many generations. Plants inherit environments that have been favorably modified by their immediate progenitors, as when certain trees and shrubs in dry habitats produce very flammable leaves, which accumulate around them to the point where they eventually ignite, burning their competitors that are not, like them, capable of surviving the flames. Then there is the extended family in many

> mammals and birds—flocks, packs and herds of related individuals, functioning as units. It might be said that these resources could not be exploited except by systems having a particular array of genes, but it is clear that these examples show that biology is much more than a molecular phenomenon, Molecular information in genes functions as switches and placeholders in a much larger material system, itself maintaining non-genetic informational constraints. (Salthe 2008, 143)

We can inherit by means of culture not only ways to perform certain activities but matter already formed (gems, oil paintings on canvas, palaces, ink printed on paper or parchment). Thus we inherit indexical signs of a life that, unlike fossils, deliberately expresses the intention to communicate. Stone carving or painting an animal also manifests the desire not only to perceive but to communicate, an effector becoming an actor towards a re-actor in semiosis. In some cases communication is an ephemeral act as in dance or singing, and in others its remains in the stone of a palace, cathedral, or pyramid, craving to reach immortality.

If we define culture in terms of artifice, no species produced so much of it as us humans, but if artifice is defined as the production of something that did not previously exist, then the production of cellulose by plants or of oxygen by bacteria could considered an artifact that far exceeds the totality of human cultural production. All human cultures produced artifacts such as pottery, scrapers, nets, baskets, picks, knives, ropes, hammers, axes, chisels, hooks and locks, necklaces, bracelets, and earrings.[17] The universality and antiquity of decoration by beads of different materials is confirmed by the findings in three continents.

As far as techniques go, fish, ants, and termites use digging and overlaying, and birds and spiders do weaving. Everything around us is artifice, the legacy of something that someone once thought, built, accomplished, and stored, and on which we totally depend through this vast division of labor that makes such a variety of devices possible. The artifact can be used and improved, changed, replicated, adapted, and developed by many generations; it can be inherited. With culture we circumvent not only extreme weather but the obstacle that prevents us from bequeathing acquired characteristics to our offspring. To protect these acquisitions, we invented the social matrices that label social territories, assign roles, and share meanings to safeguard these legacies.

The Ladies and Their Bifaces

From the Acheulian and Oldowan period at least two and a half million years ago, homo habilis and erectus, heidelbergensis, neanderthalensis, and sapiens have carved stone tools. Among our Neanderthal relatives there exist rudiments of culture, like the Châtelperronien wood and stone tools of over

100,000 years ago; harpoons to hunt bisons, horses, deer, and reindeer; or animal fur garments tied together with leather straps. Carved knives of a single edge, pointed stones, ivory jewelry, carved shells, and animal teeth have been found in Mousterian culture sites. It is unknown whether they were original discoveries and inventions or if Neanderthals learned them from Homo sapiens. Besides Oldowan stones, the first index we have of a deliberately semiotic artifact dates back 350,000 to 400,000 years in an elephant's tibia engraved with seven to fourteen parallel lines to number or mark certain significant events. (Mithen 1999)

In parallel to the development of the technical and semiotic dimension in the production of instruments, the aesthetic dimension develops through the artifice of jewelry and decorative objects as expressive masks and most probably body painting, in addition to singing and dancing. One of the mysteries of human culture is that these bifacial axes were reproduced over a million and a half years without change and in huge amounts. Symmetry is even more amazing, because being sharp in both sides, and a pointed instrument not held by a handle, they are not suitable for cutting. More amazing is that they do not apparently have a practical purpose, as charcoal 14 tests reveal they were not used for cutting or hunting. (Mithen 1999)

Mithen suggests that these instruments attest to a cultural big bang that occurred due to the evolution of neuronal flexibility allowing the integration of various types of intelligence, because these devices exhibit not only psychomotor skills but knowledge of different types of material, its strength and its location. He mentions another surprising fact: their abundance in Acheulean sites, which he explains with the hypothesis that these bifaces may have been used not just for the purpose of cutting and killing, but as indexes of fitness, good eyesight, health, skill, intelligence, and fine manual control probably made by the males to seduce the ladies. (Miller 2001, 96)

And speaking of ladies, the oldest known female figure whose classification as artifice or as random form has been hotly debated is the Venus of Tan Tan in Morocco from allegedly 500,000 to 300,000 years ago. (Bednarik 2003) Also controversial is the Venus of Birkat Ram in Israel carved by Homo erectus that would date perhaps to 230,000 years ago. (D'Errico and Nowell 2000) Ostrich eggshell beads and other decorative objects from 100,000 years ago have been found in Neanderthal sites. For Chris Knight, this autonomous development included the manufacturing and creative use of symbolic objects for the visual display of the body in Neanderthal cultures, as is often observed in various social roles reflected in traditional societies. (Knight 1996; Henshilwood et al. 2002)

There is consensus that 40,000 years ago a cultural Big Bang occurred with the production of female figures in ivory, stone, or terracotta as the Venus of Dolni Věstonice, the Venus of Lespugue, of Laussel, and of Brassempouy, or as the Venus of Hohle Fels, with exaggerated genitals and breasts carved in ivory at least 35,000 years ago.[18] The oldest cave paintings at Chauvet are dated as 32,000 years ago, the masks of the Roche-Cotard over 30,000 (by our extinct Neanderthal relatives), and the Venus of Willendorf 25,000 years ago, long before Alexandros of Antioch was able to masterfully carve the sensuous smoothness and exquisite proportions of the Venus de Milo 2,100 years ago.

Red in Ochre and Blood

These ancient and powerful Venuses show that the main topic of iconic expression in the evolution of Homo sapiens is the ladies. Moreover, it is not by mere coincidence that the Venus of Tan Tan was painted red.[19] (Bednarik 2003) According to Knight, a possible explanation is that since female papios and chimpanzees announce fertility by vivid red genitals, hominid females have no estrus, consequently menstrual blood is simulated in color as a sign of near fertility to prolong the male's proximity. (Knight 1996, 75, 80–81)

> Our focus here has been on the ritual function of cosmetics. Symbolism arose as a response to increasing levels of reproductive stress experienced by females during the rapid phase of encephalization associated with archaic *Homo sapiens*. Once reliable fertility signals had been phased out, menstrual bleeding was left as the only cue offering males positive information on which females were imminently fertile. Because pronounced menstrual bleeding was valuable for extracting mating effort from males, even noncycling females "cheated" by joining in with menstruating relatives whenever blood was flowing, painting up with red pigments to signal "imminent fertility." Dance and associated body-painting of this kind long antedates the production of representational imagery on inanimate surfaces. At the point where "collective deception" was established, such use of movement and pigments constituted symbolism. The archaeological record of ochre use in southern Africa, interpreted in the light of San and Hadza ethnographic data, matches the predictions of this model. (Knight, Power, & Watts, 1995, 96)

According to Knight et al., it was necessary to establish a code involving a collective agreement: "On this basis we would expect females, within kin coalitions, to manufacture synchrony of signals whenever a member was actually menstruating—a strategy we term 'sham menstruation.' We might then expect them to select additional cosmetics—blood-coloured pigments for use in bodypainting." (Power and Aiello 1997; Knight 1998, 80) The changing ecological variables lead to changes in mating systems between mammals through changes in females' strategies, not the males.

> Symbolic cultural evolution takes off when cosmetic displays are staged as a default—a matter of *monthly, habitual performance*, irrespective of whether any local female is actually menstruating. Once such regularity has been estab-

> lished, females have effectively created a *communal construct of "fertility" or "Blood"*—no longer dependent on its perceptible counterpart. Body-painting within groups repeatedly creates, sustains, and recreates this abstract construct. Such energetically costly repeated ritual must be linked to the level, regularity and kind of male provisioning effort it engenders. We therefore predict that data interpretable as evidence for regular female ritual performance will correlate with the onset of a symbolically structured sexual division of labour. (Knight 1996, 337–38)

We cannot emphasize enough the importance of the aesthetics in this female conspiracy as a collective shared fiction. In these women who pretend to be menstruating to keep the male around we find the elaboration of an artifice and of fiction along the lines of the theatrical poetics in the mother Kildeer bird pretending to be injured to lure the predator away from her offspring.

Following Krebs and Dawkins, "rituals emerged as conspirators' coalitions to exploit the muscle power of others who resist such a message" and add that "perhaps the most decisive advantage of a model of 'female manipulation' is that it can explain the emergence of a collective delusion that is ritual, symbolic language, and culture in general." (Krebs and Dawkins 1978, 385)

The preference bias on red is not unique to the human species, as it is found in several species, such as the case reported by Tinbergen of the aggressive reaction in sticklebacks (*Gasterosteidae*) every time they saw the red mail truck pass by the window. (Tinbergen 1939) Also recorded is the behavior of gull chicks pecking a red stain resembling red marks on their parents' peak, and the red lizard Anolis who jumped to the bowl of tomato soup in a camp, besides the curious behavior of males from the human species to red lingerie and red high heels. (Krebs and Dawkins 1978, 385) In each case, such a bias is due to a different motivation, but usually ends up obeying the same Darwinian and Biblical imperative of fruitfulness and multiplication.

There is evidence of the use of red ochre powder by the Neanderthal population as the case of a man's head covered with ochre found in Les Moustier and La Chapelle aux Saints. There is also evidence from a homo heidelbergensis near the burial site of Terra Amata south of France. Thus, we can speak of symbolic and aesthetic use of red ochre at least 110,000 years ago or more, as recorded by pigments found in Blombos Cave in South Africa.[20] Red ochre and blood in humans thus mark the beginning of an artifice whose apotheosis is exhibited today in oil-painted Madonnas, in Euripides and Shakespeare's plays, in Sor Juana Inés de la Cruz's poetry and Dante's verses as well as in Schubert and Beethoven's powerful harmonic sounds. We begin by pretending that we menstruate with the artifice of ochre and we get to express the deepest truths of human existence with the artifice of art.

BIASING

Umwelt Bias

No creature perceives everything. We are always biased by our morphology and context. Jakob von Uexküll called *umwelt* a kind of bubble that surrounds every creature and encompasses everything relevant in its environment for survival. Even at the cellular *umwelt* receptors determine what molecules can or cannot be detected, while more complex organisms are able to touch, taste, smell, see, hear to appraise and select the appropriate action.

> Perhaps it should be called a stroll into unfamiliar worlds; worlds that are strange to us but known to other creatures, manifold and varied as the animals themselves. The best time to set out on such an adventure is on a sunny day. The place, a flower-strewn meadow, humming with insects, fluttering with butterflies. Here we may glimpse the worlds of the lowly dwellers of the meadow. To do so, we must first blow, in fancy, a soap bubble around each creature to represent its own world, filled with the perceptions which it alone knows. When we ourselves then step into one of these bubbles, the familiar meadow is transformed. Many of its colorful features disappear, others no longer belong together but appear in new relationships. A new world comes into being. (von Uexküll 1992)

Von Uexküll proposed the concept of "functional cycle" or *funktionskreis* as the process of perception or reception within the *umwelt* where the stimulus or "vehicle-sign" is interpreted by the perceptor in an active response as effector. (von Uexküll 1982; Sebeok et al. 1992, 340) Such perceptor/effector operate cyclically and continuously by aesthesis and semiosis to detect, assess, and interpret environmental stimuli and act accordingly.[21] Every creature is driven by this perception-valuation of the environment from the cellular level in the amoeba's distinction between edible-inedible to activate pseudopodia, the whistling sound that alarms the animal by association to a snake, crumbs in the cornice of a London window for pigeons, or mice detecting or not the poison in the flour under the furniture.

The *umwelt* thus determines perception biases in each individual, group, and species, for quickly and effectively assessing which objects are useful and thus maximizing the limited attention and action resources. This implies that, rather than perception, what takes place is pre-ception, and rather than judgment we act by pre-judgment configured by adaptations in certain contexts. The typical example of perception bias is the case described by the ethologist Niko Tinbergen who saw gull chicks pecking red dots resembling those in the mother's beak, and reacted to any dotted red object.

William James emphasized the phenomenon of perceptual biases when he mentioned that: "The labours of Darwin and his successors are only just beginning to reveal the universal parasitism of each creature upon other special things, and the way in which each creature brings the signature of its special relations stamped on its nervous system with it upon the scene."

> [T]he nervous system of every living thing is but a bundle of predispositions to react in particular ways upon the contact of particular features of the environment. As surely as the hermit-crab's abdomen presupposes the existence of empty whelk-shells somewhere to be found, so surely do the hound's olfactories imply the existence, on the one hand, of deer's or foxes' feet, and on the other, the tendency to follow up their tracks. The neural machinery is but a hyphen between determinate arrangements of matter outside the body and determinate impulses to inhibition or discharge within its organs. When the hen sees a white oval object on the ground, she cannot leave it; she must keep upon it and return to it, until at last its transformation into a little mass of moving chirping down elicits from her machinery an entirely new set of performances. The love of man for woman, or of the human mother for her babe, our wrath at snakes and our fear of precipices, may all be described similarly, as instances of the way in which peculiarly conformed pieces of the world's furniture will fatally call forth most particular mental and bodily reactions, in advance of, and often in direct opposition to, the verdict of our deliberate reason concerning them. (James 1884)

These "peculiarly conformed pieces of the world's furniture [that] fatally call forth most particular mental and bodily reactions" would amount, according to evolutionary psychology, to the conditions that anatomically and perceptually shaped Homo sapiens. This furniture, however, is not fixed but mobile as it changes with the conditions and the niches where creatures evolve, and therefore transforms their sensory, affective, behavioral, and social structure.

Prejudices and Preferences

There are prejudices and preferences of various kinds: sensory (smell, color, tactile), landscape, tradition, play, ethnic or racial, climatic, political, religious, sexual, as each dimension and each matrix displays its own prejudices and generates preferences. Preferences obey various motives such as convenience, utility, pleasure, economy, security, custom, sometimes seemingly against survival, as the preference for self-destructive activities.

Donald Symons affirms that, being preferences, all adaptations are aesthetic because evolution has been a process that aims at strengthening perception of environment information to optimize chances of survival.

> [T]he whole notion of the "aesthetics" as a "natural" domain, i.e. as a domain that carves nature at a joint, is misguided. . . . All adaptations are aesthetic adaptations, because all adjustments interact in some way with the environment, external or internal, and prefer certain states to others. An adaptation that instantiates the rule "prefer productive habitats," is no more or less aes-

thetic than an adaptation that instantiates the rule "prefer a particular blood pressure."[22]

Symons, however, uses two different meanings of the term, first as "preference in general" or consistent tendency and in another as "interaction with the environment." There are interactions with the environment that are not preferences (i.e., penury, fear), and tendencies that are purely mechanical. He tackles the question of sexual preference in the *Evolution of Human Sexuality* and proposes that it could be the result of the chemical environment in which the brain develops during pregnancy, and of the structure of the hypothalamus, which turns much more developed in heterosexual men than in homosexual men or in females. Thus there is evidence of the biological origin of homosexuality, which in turn promotes cultural and social manifestations of homosexual sexual identity. (Symons 1979, 286–305)[23]

Among animal preferences, Robert Payne et al. studied the behavior of birds reared by foster parents and found that males sing the song of their foster parents, rather than their initial group, and females prefer males that sing the song of their foster parents. (Payne et al. 2000) Food preferences are also transmitted from mother to the rabbits and insects for certain herbs that can change depending on the ecological niche. (Bilko, Altbacker, and Hudson 1994) Humphrey says that the color preferences of monkeys are located in a blue-green spectrum so when they are in a blue room they are quiet while in a red one they get nervous. (Humphrey 1996, 202) Darwin discovered the preference phenomenon in the cabbage butterflies who lay their eggs on cabbage plants instead of others similar and equally suitable. This preference for cabbage is transmitted through non-genetic, behavioral, and environmental conditions. (Jablonka and Lamb 2005, 240)

According to M. C. Singer, a case of inherited preferences is the Euphydryas checkerspot butterfly, whose preference for laying eggs on certain types of plants seems to be hereditary. (Singer, Ng, and Thomas 1988) The challenge is to explain how such preferences are inherited. Turtles have a preference to spawn in Mazunte, Oaxaca, but are not born with a "Mazunte spawning" gene. As the tendency or capacity to learn or "learnability" proposed by Henry Brighton et al. perhaps we can talk about "preferentiality" in tuning preferences to the best options of the organism during its development. (Brighton, Kirby, and Smith 2005).

Richerson and Boyd state that "biases may be innate preferences or they may be cultural preferences acquired during an earlier episode of social learning." (Boyd and Richerson 2005, 66)

Matricial Biases

Medieval and Renaissance clergy were well aware of the irreversible and long range effect of aesthetics when they built their cathedrals, painted murals, carved sculpture, composed music, and dramatized sacramental rituals and theater to propagate Christianity throughout Europe and the rest of the world. Christianity had to show concrete benefits in order to propagate and it achieved it by the route of perceptual biases: a) *content bias* by exhibiting compassion in times of penury, as in the eloquent stories by his apostles about Jesus, b) *status bias* for advisability of joining new connections that promise greater status, better socially linked to the power of Byzantium, and c) *frequency bias* to adapt to current majorities. I will add what may be termed "antiquity bias" by the weight of tradition as a sign of fitness that inspires reverence through symbolic charges of accumulated time, emotional memory, energy. In this case, appealing to the old Hebrew Scriptures even if interpreted and translated into Greek.

More importantly, the new religion had to seduce and fascinate subjects to achieve happy evangelization. This new religious emergence set in motion one of the most impressive displays of persuasion known to the human species. The visual output of Christianity exploits human aesthetic bias in its attraction for monumentality, artificiality, solemnity, majesty, and the time and skill invested in preparing the props, scenery, costumes, music, and choreography for the ritual. In the emotion of suffering saints and inspired virgins represented in their iconography and the effects of empathetic identification with familiar scenes, these artifacts impacted with absolute mastery upon mirror neurons and human sensibility.

In short, we have a mixture of social, behavioral, and environmental aspects that condition and bias perception and action at different levels: epigenetic biases (where the daughter cell detects contextual aspects in replication), behavioral biases (behavior detection relevant to biological survival), ecological biases (detection of frequent stimuli in a given environment), matricial biases (relevant to social survival), and cultural biases (devices, languages, conventions, and techniques of a regional convention).

Fusion Bias and *Sensus Communis*

We tend not to be very tolerant of other people's preferences, as evidenced by the Kantian subject who does not ask, but demands adherence to his aesthetic judgments based on the *sensus communis*, sense in common (rather than common sense). We are greatly disappointed that someone close to us does not share our political, culinary, religious, musical, or film preferences, since by sharing preferences we bond with others or distance ourselves from them. *Sensus communis* is feeling in common with the other, longing perhaps to regain our pristine unit as in Plato's *Symposium* myth of the androgyne.

This desire to belong, to merge, probably originates in Pleistocene tribal phylogenesis and fetal ontogenesis. At a social scale, the fusion bias exponentiates among all subjects eager to be part of the masses for its intense emotional impact that propels affections and hatreds to boiling temperatures to degree of explosion. The fusion bias produces a euphoric effect of losing personal identity and being part of a human mass agglomeration oriented to a single will, as Elias Canetti (2005) describes massive meetings typical of fascist and totalitarian regimes. In fusion bias distance from others vanishes away and is compacted in a single amorphous crowd that seems to move independently and outside subjects whose individuality it annuls. Merging into the masses offers a tremendous feeling of security by frequency bias.

As plagues spread quickly in a crowded population, certain biases become more contagious in body contiguity, empathy, and multisensory stimulation, a fact that explains attraction among masses agglutinated for political, athletic, religious, or musical events. Besides the aesthetic bias used as binder like songs, posters, slogans, gestuality of the leader, his body appearance, and effect of a supraindividual mass, it simultaneously exploits other biases such as conformity, status, and frequency and, in particular, the fusion bias.

Faraday explains that cohesion is lost in water at high temperatures and increases at low temperatures when water crystallizes. Take the famous example of iron at normal temperature with impenetrable cohesion in contrast to red hot that when heated can be deformed and fragmented with relative ease. This law applied to *Tertius primus* teaches us that to deform or manipulate a group of people and make them pliable, we have to raise their emotional temperatures. In the political arena, rhetoric can heat up the crowds and their emotions to a feverish degree in an expansive social explosion. If the temperature is too low, the masses remain crystallized in inertial flow (as in hard vote). Few tactics are as effective to inflame the masses as the hatred of a scapegoat since it ignites with a simple spark.

The proliferation of slogans and icons of political leaders like Mao, Stalin, or Hitler, has exploited the frequency bias's potential to which its recipients are susceptible. Being many is interpreted as being strong. The same goes for commercials that are repeated ad nauseam on radio and television. Goebbels knew this mechanism that in order to implant into the minds of docile citizens false memes it is sufficient to repeat them thousands of times until their cerebral cortex is tattooed with the hot iron of such repetition. One such sign was the swastika, chillingly successful as a simple black axed scheme rotating in a white circle on a red background. It exploited the red bias and the concentric circle bias prevalent among animals, together with clockwise mechanized revolving axes particularly appealing to sadomasochistic re-actors' tendencies that require order, discipline, toughness, monotony, and a static, destructive, and controlled movement. Fusion bias is also exploited by the impact of hypnotic graphic signs as the sickle and hammer, the singing of hymns as *Deutschland, Deutschland über alles*, *La Marseillaise*, or *Horst Wessel Lied* sung in unison by the crowd and the infectious, rhythmic slogans as *Sieg Heil* (hail victory), "*Este puño sí se ve*" (this fist can be seen) and the "*Zadom Spremn!* (ready for homeland!) in Croatian fascism. We should add the somatic mass gatherings such as rock concerts, massive religious and state assemblies, particularly the impressive performances designed by Speer for the Nazi party in the Zeppelin field. On many occasions art has recruited fusion bias to exploit emotional energy affected by resentment and frustration and has created propaganda through utopic, melodramatic, or manicheistic narratives that engage susceptible receptors.

Sensory Exploitation: A Cynical Perspective

The limitation of vital resources in the body and the fierce competition among cultural variants of mental hosts to replicate significantly increases the attention to the role and effectiveness of aesthetics. Cultural and subcultural traits emerge based on very superficial aesthetic differences for their efficiency in the transmission of messages to capture adherents. Their most perspicuous function is to mark cultural niches for establishing identities, as in the case of contemporary urban tribes such as emos, punks, goths, metalheads, floggers, chetos, glam, and others that require easily decodable signaling to find or to reject relationships in the context of hard urban anonymity.

The advertising industry is the absolute master in exploiting perceptual biases. The basic mechanism it uses is the association of an item as bonus with another operating as bait: an attractive model as bait for sexual bias and a car (or vice versa) as bonus, the bias for red as bait and the coke as bonus, the bias for blue as relaxation to associate with an airline flight as bonus. Psychological biases are also exploited, typically the bait of maternal guilt or a woman's need for tenderness by buying the bonus of a fabric softener. Obsessive-compulsive biases and sexual phobias can be exploited, advertising bacteria killers as bait and soap as bonus. I remember the case of subliminal seduction advertising with sexual imagery as bait (in print and filmed), where the barely distinguishable word "sex" in a few ice cubes of a gin glass sexually aroused the viewer to buy alcohol as bonus. (Packard 1991)

John Krebs and Richard Dawkins (1984) argue that the most effective features in advertising are redundancy, rhythmic repetition, flashy packaging, conspicuity, contrast, and supernormal stimuli (exaggeration) that are found in many animal world signals. (Sforza, Feldman, and Chen 1982)[24] They defined a "sign" as "anything that acts as an incitement to action" and explained that these signals can be of attraction, repulsion, or transformation. They assume a position they call "cynical" by viewing a signal as the means by which one animal (the "actor") exploits another animal's muscle power (the "re-actor's").

> Many of the externally visible features of animals, many of their behavior patterns, many chemical substances and most of the sounds given off by them, are best interpreted as being adapted—designed by natural selection "to influence the behavior of other animals and are often referred to as signals" (Some authors reserve for "signals" to morphological features like crests, and use displays for behavior patterns). Just as a wing performs its normal function by working on the air, so a signal performs its normal function by working on another animal via its sense organs. (Krebs and Dawkins 1984, 380–1)

Under this proposal, we can understand that the cricket sends acoustic signals to attract a female to use her body for replicating his genes. Such signals may even influence the physiology of another animal as the canary song that accelerates the maturation of the female's ovaries. (Hinde 1970) For cynical neo-Darwinism, the actor's signal is emitted to its own species or to other species' re-actors, as acoustic mimicry of some birds that imitate the hissing of snakes to scare predators.

Cynical Darwinism proposes a co-evolution of roles between what they call "manipulative" or issuer (actor) and "mind reader" or receiver (re-actor) that alter the behavior for their own benefit. If the benefit is mutual, signals are economized and if only one party benefits, the signals become very conspicuous, repetitive, and ritualized.[25] But if the interests of the sender and receiver are very different, selection favors resistance and acts on issuers or actors to overcome this resistance, promoting the evolution of conspicuous and complex signals.

> An intriguing idea is that the kind of signals that evolve depends on the degree of common interest between the sender and the receiver in an interaction (Krebs and Dawkins 1984). Two co-evolutionary pathways are envisaged. If there is a high degree of common interest between sender and receiver, selection acts on receivers to become more sensitive to signals and on signals to become simple and inconspicuous (i.e., cryptic). But if sender and receiver interests differ, selection favours 'resistance' in receivers and acts on senders to overcome receiver resistance, promoting the evolution of complex, conspicuous signals (See also Williams 1966). (Arak and Enquist, 1995)

We recall the peacock case in which, as if its color and the shape and size of its tail were not enough, the male moreover flaps his feathers drumming them whenever a female approaches. According to this explanation, the peacock's tail would result from the lack of sensitivity of the peahen. He would thus require deploying that portentous tail to get her attention. Perhaps it is a matter of not getting a particular benefit to mate with that particular male, being polygamous and not contributing to the breeding of offspring. He must then advertise himself harder than other males to overcome female resistance. Finally, it can also be an index of the female's very fine sensitivity to discriminate more sharply the quality of a prospect by the feathers' symmetry, color intensity, concentricity, flexibility, and firmness.

If we project Dawkins and Krebs's cynical perspective to art, we can consider Warhol's serial portraits of famous people, Lichtenstein's vignettes amplifications, and most action painting's huge canvases, so conspicuous in their exaggerated color and size, to affect not very interested art spectators or as works too eager to compete against others in the art market for grabbing not very sensitive collectors' attention. Pollock's canvases can be interpreted as sensory signals to exploit the viewer, in this case Mr. Smith a potential collector (and force him to give away dollars he painfully earned in his boring office job enduring stress and his boss's abuse, after having suffered grueling exams and classes for years of a soporific business career). All of this so that Pollock can enjoy a good life in his Soho loft, getting up when he pleases, drinking booze and dripping paint. Maybe Smith can feel some relief to see Pollock's painting in his living room, or perhaps he just uses it to display the acquired famous artwork at his wife's organized dinners to promote him and increase his status within the company.

That advertising and propaganda exploit these biases does not mean that we are helpless. It is possible to resist since the re-actor interprets and reconstructs the stimulus and can ignore, mute, or filter it by modulating the *tertius* threshold which otherwise would be reduced to automated responses to these biases. The starting point is to recognize that such biases exist although matrix pressures can operate as high power loudspeakers to reinforce such biases. Matrices' dynamics tend to generate and reproduce these biases and preferences since the individual almost never thinks alone. Marx referred to this malleability when speaking of the determination of consciousness by the material conditions of reality, namely the set of biases that result from matricial location. Hence, the best corrective to adjust and refine perceptions can be traffic between matrices to relativize their biases.

The Aesthetic Bias

In the historical development of cultures not only technology, political and legal systems, or scientific paradigms have undergone significant transformations. What Norbert Elias called "social habitus" imposes prohibitions and requirements of an aesthetic character that have changed, categorizing some attitudes as pleasant and appropriate while others as repulsive and inappropriate. (Elias 2000, xiii) The coarse, the delicate, the disgusting, the shameful are aesthetic values and judgments dependent on culture that end up affecting, refining, or destroying sensibilities. Elias illustrates his point with the fashion of taking a bath with the help and presence of the servants during the fifteenth century or urinating in the streets of Paris. During the colonial period in the capital of New Spain it was customary to throw the contents of chamber pots from the windows to the streets of the historic center shouting "waters!" out of consideration for the potential unfortunate pedestrian.

Given the relativity of fashion and manners, of what is appropriate and inappropriate, a theme that inspires Hume's essay on the Standard of Taste, Kant built with admirable tenacity his *Critique of Judgment*, massive theoretical architecture in order to establish foundations for the objectivity of aesthetic judgment. (Kant 1790; Hume 1965) Ironically, the objectivity that Kant sought would finally be found (oh surprise!) not in judgment but in prejudgment or aesthetic pre-judice.

The aesthetic prejudice or bias is the tendency to perceive, assess, and highlight certain aspects and patterns as valuable through the amorphous mass of the environment's stimuli and discard or ignore others. This bias depends on adaptations that have been inherited by natural and social selection throughout our evolution and acculturation. Whoever does not react immediately to snakeskin, bright color of ripe fruit, glow of dew upon a leaf, the symmetry of a face looking straight at us, could die from biting, hunger, thirst, or devoured by a predator.

Aesthesis processes begin from molecular and cellular receptor couplings to nervous systems with complex sensorimotor terminals. Passing the threshold of *tertius*, sensitivity diversifies exponentially to attend social facets and cognitive techniques that obey very different codes. Several authors have studied perceptual biases in different species and observed trends that respond to certain stimuli: more intense colors or signals in the domestic fowl, longer fins in fish, more symmetrical peacock tails, and so on.[26] I mentioned the case of satin bower birds or passerine birds with bias for blue to decorate their love nests. Humans are apparently also biased towards blue in landscape paintings as shown in taste polls by Komar and Melamid (1999).

The case of the Congo chimp—who avoids mixing colors when painting—was quoted by Desmond Morris and later corroborated by Eibl-Eibesfeldt with the Munich zoo chimpanzees.[27] He also noticed that a female chimpanzee painted curves and parallel bands like a rainbow instead of saturating all colors into brown. Both cases appear to show that there could be a bias to separate color spaces which may be a result of cognitive and practical benefits of distinguishing objects and highlighting figure from background.

Evolutionary psychology proposes that several animal species, vulnerable to snakes, have evolved an innate system adaptation or specialized module to detect them. These innate biases are illustrated in cases of many small mammals' innate fear of snakes' hissing noises or of hyenas that have never seen a lion but were scared the first time they detected its smell. From this system Richard Coss (2003) derived the idea of bias that is expressed in our attraction to the tessellated shapes as a psychological adaptation of immediately recognizing snake or crocodile scales. Perhaps that explains our attraction to iterative zigzag patterns in decorative objects like kilims, mosaics, tapestries, embroidery, fabrics, carvings, and jewelry so widespread among the most diverse cultures.[28] I just realize at this moment that the clay floor tiles at my home actually look like wet scales, and that is perhaps why I chose them and enjoy looking at them. Apparently, there is also an aesthetic bias for leopard spots and black rosettes on yellow background, typically preferred to express sexual availability.

The attraction for shining metals like gold, silver, bronze, and copper is probably related to the importance of detecting light and water, and may partly explain our fascination with diamonds. The bias for surface brightness may have involved the difference between life and death as in wet dewdrops on leaves, and hence may derive the tendency of young children to lick shiny surfaces. Coss (2003, 87, 72) examines phenomena designated as "cognitive bias" and notes that natural selection certainly operates on visual tasks that impact fitness. He refers to four operations: detection, recognition, meaning, and action (basically, von Uexküll's *funktionskreis*, since the bias's task is precisely to economize the detection-recognition-meaning effort).

In this line, Christa Sütterlin states that "*all* perception is biased" because that's how our system works and wonders whether sensory natural perception operates similarly to aesthetic perception. (Sütterlin 2003, 133–34) However, as I have been arguing, aesthesis starts as perception, and obviously depends on it. "Thus, perception does not mirror the environment in a mathematically correct copy, point by point, but scans first for what is regular and repeated, i.e. statistically likely to occur. This saves memory in the brain and helps to organize the flow of information in the visual field." (Sütterlin 2003, 166)

Preconscious scanning of features biases every act of perception. "As on the purely formal level of *Gestalt*, these kinds of prejudices act as visual codes or guidelines in channeling information towards a certain prototype/archetype of internal representation and influence the formation of aesthetic preferences as well." (Sütterlin 2003, 167) She moreover assumes that painting may have been initiated by the abstraction attributed to Neolithic period outlines of perceptual experience 9,500 years ago rather than by naturalism in Paleolithic painting as in Altamira and Lascaux. In the Neolithic, a variety of objects were created such as vessels of baked terracotta, ceramic figurines and baskets woven and decorated with figurative images or geometric shapes. Hunting and fighting weapons were manufactured during the copper, bronze, and iron ages (like Yetholm copper shields 1200–800 BCE). It must be emphasized that the goal here was to achieve not only useful objects but decorative effects by filigree, damask, embossed metals or leather, and so on. In the decoration of these artifacts, the aesthetic biases mentioned are saliently expressed such as tessellated shapes, zigzag designs, frets, color separations in parallel bands, and reliefs imitating scales.

Our aesthetic bias towards concentric circles apparently arose from the need to detect the eye's iris and pupil of someone observing us. Circular shapes are present in various animal species decorations as ladybirds or ladybugs, monarch butterfly wings, peacock feathers, or spots resembling eyes in fish like *Paralichthys albigutta*. This bias toward concentric circles has interesting consequences, as James Lennon proved by an experiment in a department store where he placed transparent panels with a simplified version of eyes, which dramatically reduced the theft of merchandise (as if being watched), but not the sale. Hence archaic biases are expressed today in decorative crafts among all peoples, for example concentric circles as eyes, wavy lines as water, spikes, zigzags, floral motifs, alliterative patterns at different scales of the type of fern leaves, fractal structures like broccoli, or pearl necklaces that are a fascinating analogy to perfect teeth.

There is attraction to plants such as yucca or Spanish bayonet glorious (Spanish dagger) by its sharp, elongated leaves, which have derived from the fact both of their usefulness as dagger-like instrument and their risk. This bias is continually expressed in multiple motifs of indigenous cultures in spikes, picks, and tips of incisors and exemplified by the aggressive design of the chapel of the Academy of the U.S. Air Force in Colorado Springs by Walter Netsch, as noted by Coss (2003).

Aesthetic prejudices common to the human species highlight aspects such as balance, symmetry, brightness, rarity, intensity in color or gesture, saturation, focus, concentricity. The practical evolutionary reason that explains stability and balance is that they express security; intense saturated color is an index of vitality and health. Symmetry indicates resistance to parasites (can be smelled by lizards and humans or certain scorpion flies), and the preference for males with more symmetrical feathers among barn swallows (*Hirundo rustic*) and many other species has been clearly observed. (Hamilton and Zuk 1982; Thornhill 2003, 18)

Dondis (1976, 28–9) proposed categories for visual syntax such as regularity and irregularity, dynamism and stasis, continuity or randomness, predictability and spontaneity, balance or instability, symmetry or asymmetry, variation or consistency, sequentiality, passivity or activity. As Pinker (1997, 536) would point out in terms of cognitive psychology, we generally exhibit preferences for contrast, sharpness, tone, accent, boldness, focus, and brightness or for strong, clear, analyzable signals rather than distortion, fuzziness, vagueness, ambiguity, indecision, and neutrality.

The big-nosed monkey Nasosus Machedonus (*lavatus nasalis*), who hangs his nose outside against his mouth, attracts the female by his seductive double bass voice. If the female were to judge by the sight, would she fall into his arms so lovingly? It all depends on the bias of the beholder.

NOTES

1. I wish to thank Editorial Siglo Veintiuno Editores for permission to reproduce as epigraph Borges's phrase taken from the short story "Tlön Uqbar y Orbis Tertius" in *Nueva Antología Personal*. 2000. pg. 80.

2. This interrelation between the outer world and the brain is what Roger Bartra called "exo-cerebrum." (Bartra 2007).

3. Cf. chapter 2, fourth section, on these categories.

4. Exceptions are rare: Philippino ina, Galleho nai, Hungarian anya, Indonesian and Malay ibu, Turkish anne.

5. On the origin of language cf. Mithen (1999, 203, 255).

6. Two last ones taken from Parret (1995; 1993).

7. Cf. Gallese et al. (1996), Rizzolatti (2004); Dinstein et al. (2008); Iacoboni et al. (2005); Prather et al. (2008); Brainard and Doupe (2000).

8. Dealing with this subject cf. Ritchie (2000); Jakob, McClintock, Zelano, and Ober, (2002).

9. On the concept of the formation of cultural matrices cf. Mandoki (2007, Part VI.)

10. Extensively explained in (Mandoki 2007, Part VI).

11. On the inequality of place, I have argued amply cf. Mandoki (1998).

12. Tinbergen (1939); Krebs (1971) quoted by Wilson (1980, 35).

13. Johnsgard (1967) in Wilson (1980, 142–144).

14. Cf. Cockburn, Cynthia and Ann Oakley. "The culture of masculinity costs all too much to ignore." *Guardian*, November 25, 2011. Accessed November 27, 2011; British Crime Survey. http://www.homeoffice.gov.uk/science–research/research–statistics/crime/crime–statistics /british–crime–survey/. Accessed November 27, 2011.

15. Quoted by Cockburn and Oakley (2011).

16. "A Mexican Cartel's Swift and Grisly Climb," *Washington Post*, June 13, 2009.http:// www.washingtonpost.com/wp-dyn/content/article/2009/06/12/AR2009061203829.html? sid=ST2009061204134. "La estética del narcotráfico," *Esferapública*, June 28, 2009. http:// www.offnews.info/verArticulo.php?contenidoID=21372. Accessed November 27, 2011.

17. Mentioned in Pinker (1997, 189).

18. On female figurines see (Conard 2009; Henderson 2009).

19. By termoluninicence on burnt stones it can be established, a date average of 77,000 years for the layers containing ochre engravings.

20. Wreschner (1980) and Philip and Dibble (1987) suggest that menstrual blood is the source of such demonstration.

21. On the relation aesthesis-semiosis cf. Mandoki (2004; 2007).

22. Quoted by Thornhill (2003, 9).

23. On the biological substrate of sexual orientation see also LeVay (1991).

24. Cavalli-Sforza, L. L., M. W. Feldman, and K. H. Chen. "Theory and Observation in Cultural Transmission." *Science*, 1982.

25. See also Williams (1966); Arak and Enquist (1995).

26. I am referring to Jansson and Enquist (2003); Endler (1998); Arak and Enquist (1995); Krebs and Dawkins (1978); Dugatkin (1996); Zahavi (1975).

27. Morris (1967); Eibl-Eibesfeldt (1966) quoted by Sütterlin (2003, 141).

28. Ramakrishnan, et al. (2005) discovered that monkeys differentially detect different types of snakes and can discriminate between poisonous and harmless.

Chapter Five

Excess

μηδὲν ἄγαν

"Nothing in excess" (*Mēden ágan*), warns the maxim inscribed on the walls of the temple of Apollo at Delphi. Nothing except Dionysus, instigator of every kind of excess: the mystic, the orgiastic, and the ecstatic ritual dances, the euphoric and intoxicating revelry in bacchanalia. Dionysus was the patron of wine and of the most passionate arts like singing, drama, dance, and poetry, all driven away from Plato's Republic precisely because of it. Dionysus's symbolic presence invokes the sense of freedom, fertility, generosity, and joy.

In *The Birth of Tragedy*, Nietzsche writes that art is born of the union between Dionysian and Apollinian forces. The Apollinian is symmetry, proportion, harmony, and order was the canon for poetics. Apollo rules composition of classical architecture and sculpture in Greece and Rome. We find the Apollinian in the sober medieval religious representations and in Renaissance and neoclassic paintings, in Bauhaus formal rigor, in Malevich's Mystic Suprematism and Mondrian's Neoplasticism, in Lucio Fontana's fine Spatialism, Goeritz's Minimalism and Hersua's Geometrism, in Rothko's spatial spiritualism and, especially, in Bach's perfect harmony and God turned music. But Apollinian aesthetics also may punish for yielding spontaneity away in its eagerness for formal purity.

Dionysian excess, on the other hand, is present in the intricate work of Chartres' stained glass, in the overflowing imagination of Arcimboldo, in the plethora of forms at the Alhambra, in the patriotic intensity of Chopin's Polonaise, in the cornucopia of images in Bosch's *Garden of Delights*, in Tonantzintla's reliefs of angels, in Munch's terror, in Modigliani's languor, Beethoven's impetus, and in Heavy Metal's acoustic energy. The impact of gothic, baroque, rococo architecture is due to their excessive scale, abundance, strength, depth, fervor. Apollo and Dionysus's aesthetics are opposites in restraint and excess, yet both have deep evolutionary roots.

Every culture produces material excess which can be dealt with in four possible ways: accumulation, destruction, distribution, and dissipation. Accumulation is the norm in tributary and slavery systems, as well as in affluent layers of capitalist society. In destruction, such as dumping, goods are secretly discarded away to the garbage. Distribution is obviously the most decent and ethical manner to deal with surplus in a context of social inequality. However, anthropologists have found hardly any evidence of equitable distribution of resources either in Western or in non-Western cultures, while the other three ways of dealing with the excess are frequent. Why?

Dissipation is the most widespread form of managing excess and, oddly, it is always exhibited and performed in public. There may be a shortage of resources in some Mexican barrios of Xochimilco and Iztapalapa or among the Tlingit and Haida tribes in Alaska, but the excessive expenditure and display of fiestas del santo patrono by mayordomos or by chiefs during potlatch ceremonies are obdurate traditions.[1] The basic questions to be asked are: why is excess dissipation so widespread, much more than its distribution, among quite different cultures? And why is the aesthetic dimension always associated with these excessive festivities in the *tertius*?

Four authors have treated the idea of material excess but mentioned aesthetics only tangentially: Thorstein Veblen, Marcel Mauss, Georges Bataille, and Amos Zahavi. To explore the relationship between aesthetics and excess in both *Orbis secundus* and *tertius* we must distinguish two different approaches: the *aesthetics of excess* in art, nature, and everyday excessive objects and *aesthetics as excess* in evolution that so intrigued Darwin. We will examine the natural and social processes by which aesthetics turns into a manifestation of excess and excess into aesthetic expression through a variety of concretizations among the most diverse creatures and cultures.[2]

THE NECESSARY AND THE SUPERFLUOUS

There are two ways to demarcate the idea of the "necessary": the logical or philosophical sense (as opposed to the contingent or random), and the functional or practical sense (as opposed to the superfluous). Adam Smith defined the socially necessary in the second sense:

> By necessaries I understand, not only the commodities which are indispensably necessary for the support of life, but whatever the custom of the country renders it indecent for creditable people, even of the lowest order, to be without. A linen shirt, for example, is, strictly speaking, not a necessary of life. . . . But in the present times, through the greater part of Europe, a creditable day-

> labourer would be ashamed to appear in public without a linen shirt. . . . Custom, in the same manner, has rendered leather shoes a necessary of life in England. The poorest creditable person of either sex would be ashamed to appear in public without them. . . . Under necessaries, therefore, I comprehend, not only those things which nature, but those things which the established rules of decency have rendered necessary to the lowest rank of people. (Smith, 1776, part 2, article IV)).

For Smith there are needs (in the *secundus*) "for the maintenance of life" and needs (in *tertius*) for what "renders it indecent for creditable people, even of the lowest order, to be without." Who are the "creditable people" and to which "lower order" it concerns are issues to discuss, but what is clear is that matrices in the *tertius* define what people need in order to be socially credited.

Georges Bataille (1987) proposed a truly Copernican revolution for the view of economy based on the concept of excess rather than scarcity. In contrast to the dominant paradigms of struggle for survival in competition for limited resources, Bataille argues that both *secundus* and *tertius* obey a pattern of excess and abundance, not precariousness. He notes that a living organism receives more energy than it needs and that this excess is not only unavoidable but must be dispersed or else it may become destructive. Excess sperm (120 to 600 million) for a single ovule, the excess of ova deposited by many species, the excess of female *jouissance* or sexual pleasure that Lacan had such difficulty understanding, all show a tendency to dissipation and exuberance.[3] We should add the dramatic fact of the 6000 eggs laid by the Pacific female salmon, all of which but one or two must perish. (Trivers 1985, 12)

For Bataille, twentieth-century World Wars were the catastrophic consequence of industrial excess that failed to be dissipated or distributed in time. Today's consumer societies, oligarchies and the concentration of wealth by a few families and penury to the majority of populations illustrate the point in almost every country.

Bataille switches from the functional to the logical sense when he designates the economy of excess as a non-contingent process, like the sun which necessarily produces an excess of energy. However, this excess in the *primus* is necessary, not casual. Superfluity is a relative value that depends on an anthropocentric view calculating utility or benefit for a specific purpose and a specific species. Aesthetics mediates, as we shall see, between this logical sense of excess being non-contingent—as can be found in all cultures, no matter how limited their resources—and the functional sense of excess as a necessity for dissipation and group cohesion.

Excess in *Orbis Secundus*

The vastness of the *orbis primus* is unimaginable, too excessive to conceive or imagine or reason even through Kant's concept of mathematical sublime. Such universe is, furthermore, expanding, and could well be just as a quark of another unknown universe. In the *orbis secundus*, excess is linked to aesthetics, typically the peacock's astonishing tail. One of the most interesting hypotheses to clarify this excess in nature is Ronald Fisher's hypothesis of "runaway process" explaining the origin of these exaggerated characters by reproducing not only such a trait but the preference for it. In other words, the next generation males will inherit that trait while the next generation daughters will inherit preference for that trait in a process that becomes exponential. For Fisher this tail is merely a fad or whim to please the taste of females. (Fisher 1999)

Amos Zahavi proposes another, more practical explanation for this aesthetic excess in peacocks through the "handicap principle" hypothesis. To emit a signal it is necessary to back its authenticity by paying a cost in exhibiting it as excess, namely, putting the emitter at a disadvantage to prove that such a signal is honest. To prevent cheaters from imitating them, individual quality signals should not be easily faked. (Zahavi and Zahavi 1997) Moreover, if there did not exist animal cheating and their capability for lying and producing illusory or false impressions, there would be no handicap principle. In other words, even if not every animal behavior may be understood according to Zahavi's handicap principle, the fact that it exists proves that there is an ability to deceive and construct illusions or fiction by different species.

But there is more. The most spectacular and colorful feathers apparently occur in the species most vulnerable to parasites. Brightness and color indicate an energy investment, according to W. D. Hamilton, which would be reduced in parasite infested animals in direct proportion to their vulnerability. Consequently, since diseases or parasites alter the symmetry and color of their feathers, by showing its tail, the peacock also exposes his complete medical record of immunity or vulnerability. (Hamilton and Zuk 1982) This hypothesis places the aesthetic phenomena in its semiotic and pragmatic framework and brings into question the false dilemma or opposition between the useful and the beautiful. Beauty can be useful or useless, it can be gratuitous and costly, sophisticated and primal, excessive and indispensable.

In addition to the peacock's tail, another peculiar phenomenon of excess observed by Zahavi in nature takes place by overspending energy and putting oneself at a disadvantage to dissuade the predator, as the gazelle leaping to draw attention from the predator and exposing herself to danger. Zahavi proposed that these behaviors are selected in the evolution of the species for symbolizing in their very excess their carrier's quality. Moreover, excessive characters diverted from the norm are discarded by natural selection which tends to remain in the middle range. (Trivers 1985, 22) Choice, however, sometimes also acts in quite disruptive manners favoring extremes and splitting species.

It is very expensive to hold up each statement on the symbolic order because symbols depend on an investment of matter, time or energy. Hence, the emergence of the signic order because it operates more efficiently simply by differences and oppositions in a system of conventions, rather than through the symbolic order that requires a material or durational backing.[4] Zahavi therefore emphasizes the importance of standardization to evaluate a sign providing a common measure for better discerning the differences.

The paradise bird hanging upside down on a branch, flapping its wings, spreading its feathers to impress the female illustrates this excess energy, and material cost of symbolization as an index of quality. Nevertheless, it also exposes it to display its failure.

Zahavi distinguishes between: a) traits evolutionarily selected for their usefulness to survival and b) signals that are selected for putting the emitter at a disadvantage to guarantee its reliability or honesty. For example, a long nose indicating the orientation of the gaze is a handicap giving away the animal's attention and intentions. Another example is a longer fragile neck that weakens the bearer in battle, but at the same time indicates its ability to cover a greater range of view in the horizon. It may be useful to distinguish between *features* that have use value and *signals* with exchange value, where the features operate through the perceptor/effector cycle while signals are directed to someone and require the actor/re-actor (as mentioned in the distinction of the three thresholds of perception in chapter 3).

The display of excess, like all aesthetic manifestations, unfolds through the four registers previously mentioned: the somatic in acts that require vigor and are risky like antics, or in the chemical-molecular discharges of attractive or repulsive smells; the scopic or visual appearance, color and decoration; the acoustic in vocalization, singing, or stridulation; and the lexic by exhibiting dexterity in a special use of language in the human species. Somatic display is exemplified by larks that do not open their wings in flight until the very last minute to show, by such excess, their quality and skill. The beta fish or Siamese fighter keeps his fin upright during a fight to display its disadvantage in breathing and therefore the cost of energy and fitness. In the scopic, the bowerbird exhibits an excessively decorated love nest with bright colors and rare items and even paints them with berry juice. The black wheatear (*Oenanthe leucura*) collects a kilo and a half of stones into its nest to impress the female.[5]

How to account for a behavior as excessive as Menura lyrebirds with their complex singing and lyre-shaped tail? Why does the mockingbird sing for hours without anyone listening except himself? A song has a great metabolic cost, as proven in the nightingales that lose weight after a night of singing. (Rothenberg 2006, 33, 34, 41) A synchronized singing quartet between two females and two males discovered by Peter Slater in South America rejects any practical explanation except, perhaps, bonding. Rothenberg thus considers that excess in itself could be a kind of evolutionary message. (Rothenberg 2006, 39, 234, 127)

For Umberto Eco, "semiotics is in principle the discipline studying everything which can be used in order to lie." (Eco 1976, 31) If animals did not lie or to produce illusory effects on perception, such excessive spending in signaling and crediting as we find by the handicap principle to prove signals' authenticity would not have evolved. The human species may express its imagination and fantasy through film or theater, literature, painting, and sculpture, but we are not able to change the color of our body like octopus, fish, poison dart frogs or chameleons by chromatophores imitating another species to deceive a predator.

Excess in *Orbis Tertius*

Zahavi extends the logic of the handicap principle (which always involves an excess) to the *tertius,* where he will meet Thorstein Veblen (1924) and his theory of conspicuous consumption. The old Chinese tradition of tying girls' feet to exhibit docility, domesticity, and leisure in women by handicapping their mobility, and the custom of growing long nails as a handicap for performing manual work illustrate status display and the ostentation of leisure. The current use of very high heels exhibits the handicap of sacrificing comfort to indicate the outlines and strength of the leg, emphasize women's hip movement and agility, youth and energy as well as their availability and desire to attract males. The recent fashion of bare belly blouses or women's pelvic level pants exhibit perfect metabolism and food self-discipline of the carrier, as tattoos exhibit the ability to endure pain as well as pleasure. Exhibiting shoulder width enhanced by a tie as index of masculinity functions by the handicap principle. Bullfights exhibit the risk of the bullfighter's genitals as a display of masculinity, cruelty, and endurance in a game of life and death through the handicap principle.

Zahavi deduces the origin of art and crafts by face decorations that emphasize qualities and conceal defects. Decorating a round vase with a circle exhibits its perfect roundness and regularity of the contours. It also emphasizes the psychomotor ability of the artisan to observe and execute, as the case of bifaces. The old-fashioned European dowry exhibits the patience, dexterity, and fine manual control of the bride to be, and signals her sensitivity and suitability to become a mother, traditional virtues for parenting and home care. All this is an excess.

AESTHETICS OF EXUBERANCE

Exhuberance is beauty
—Blake[6]

The idea that excess has aesthetic value seems to contradict mainstream theory of art tenets for good taste and high quality art because it is against the principle of economy of means for its indulgence in the superfluous. However, no one can remain indifferent to the aesthetic manifestations of excess in the *secundus* or *tertius:* lush in vegetation, the abundance of stalactites and stalagmites in the of Postojna and Mammoth caves and the colossal scale of the Hypostyle at Karnak Temple, the overwhelming power of Niagara Falls or the Devil's Throat at Iguazu Falls, the monumental expansion of the Great Wall of China, the dazzle of the Hall of Mirrors at the Palace of Versailles or the opulence of the room of lights at Catherine I's Palace in St. Petersburg. The prototype of the most extreme excess is the *mexuar* in the Nasrid palace at the Alhambra, absolutely excessive ornamentation of ceramic tiled geometric motifs on plasterwork and profuse decorations of lintels and *mucarna.*

Aesthetics in the *tertius* swings as a pendulum between moderation and excess. In the artistic matrix incorporeal bodies of medieval painting materialize voluptuously during the Renaissance with abundant carnality, particularly by Raphael, Leonardo, Michelangelo, and later in Goya, Velázquez, Courbet, Ingres, and, of course, Rubens's sensualism. The formal severity and contained emotions in Jean Louis David's paintings contrast with the ardor of Delacroix and Gericault's romanticist scenes. High strung emotion overpowers Parmigianino's Madonna in her elongated neck, Greco and his languored monks, or Goltzius's *Icarus* stunning corporeality.

The formal excesses that thrived during the Art Nouveau period were abolished by Bauhaus's functionalism and puritanism to reappear again and turn to postmodernism's chromaticism, ornamentation, and humorous allusions, quotations, and intentional anachronisms. Minimalism, geometricism, concretism, suprematism, neoplasticism, and spatialism achieved formal control to minimum standards (Malevich's white square on white background could not go further in formal and chromatic austerity). Across the pendulum, excess in color and emotions is allowed in expressionism and fauvism, exaggerated onyric figures of surrealism, and Dada's lucid irrationalism.

In contemporary art, signs of excess are equally impressive: the exaggerated scale in Christo's concealment works, the exaggerations of mass culture in pop art as Lichtenstein's cartoon amplifications, Oldenburg's huge inflatable devices, Warhol's strident color and alliteration, Jeff Koons's giant photos, or action painting's huge canvases. It should be noted as well the stubbornness with excess in the viscerality of Punk art, body art, and performance, as Orlan's grotesque facial transformations, Stelarc's grafting a third ear on his arm, Ulay and Abramovic's self-imposed bodily pain, or Stewart Brisley's rituals (*Art for Today*; *Nothing* 1972). For his performance on August 24, 1998 at the Prinzendori castle, Hermann Nitsch used 1,000 liters of animal blood, 13,000 liters of wine, 1,000 kilos of grapes, 1,000 kilos of tomatoes, 2,000 flowers, 10,000 kilos of linen, and so on. We may question its artistic value and economic waste, but its display of excess is unequivocal.

Regardless of the categories involved (beauty, ugliness, magnificence, or grotesqueness) excess is linked to the aesthetic in that it exploits a bias and captures attention, involves our sensitivity and imagination not only in art and poetics but in prosaics of everyday life. The extremely narrow waists, bulky breasts, and staggeringly high heels are exhibited as aesthetic qualities of femininity. Excess in food by serving a whole turkey on Thanksgiving or Christmas dinner for a nuclear family is seen as aesthetic expression of abundance. Excess in speeding on the roller coaster and car racing, excess of sound and volume in discos, excess of sun, sand, and sea at the beach are enjoyed by many. Thousands of crashed cars, of burned buildings, and of endless explosions represented in film and television are there to satisfy the appetite for the spectacle of excess and destruction.

The excess of space in monumental architecture especially in totalitarian, despotic, and autocratic regimes, of the physical effort exhibited by athletes, of absolutely exaggerated costumes in fashion shows like Jean Paul Gaultier's, of drag queens at contests, of necklaces and ornaments at New Orleans Mardi Gras, of costumes and dance at Rio de Janeiro carnival are measured in terms of their aesthetic effects of abundance and excess.

Porcelain figurines, stuffed animals, and souvenirs are displayed as relative excess in precarious conditions equivalent to the exhibition of art works among wealthy classes. Gifts are the prototypical symbols of excess: jewels are gleaming, perfumes are pleasant, liquor is luscious, bouquets are lovely, chocolates delicious, and bonsai cute; none is necessary, all are excessive and each is aesthetic. We may react with pleasure or displeasure to the excessive, but we can never remain indifferent to it. Excess is never aesthetically neutral.

THE *HAU* AND *AURA* OF THINGS

Mauss initiated an economic inquiry and ended up with an investigation into morality. He wanted to understand the code behind mandatory reciprocity: "What rule of legality and self-interest, in societies of a backward or archaic type, compels the gift that has been received to be obligatorily reciprocated? What power resides in the object given that causes its recipient to pay it back?" (Mauss 1990, 3) Remarkably, Mauss implies in the second question ("what power resides") a partial answer to the first: It is the belief that there is a power within objects that acts upon people and forces them to reciprocate gifts. This power is the *hau* or spirit of objects, which retain part of the soul of their maker and can be destructive if not properly recognized and dealt with. The Maori people call *hau* this spirit that clings to an object when ownership changes: one must relate to this concrete presence in objects when introducing them into one's home. (Mauss 1990, 11)

Mauss explains that *hau* exists also in personal property as well as in nature and that it influences people's actions. From a contemporary point of view, the idea of *hau* seems mere superstition. Nevertheless, consider that our culture strongly forbids giving away a gift received from someone else, as if the spirit of the gift giver would be betrayed in doing so. If it were just an object, such a ban would have no basis, but the gift seems to keep coagulated energy not only of the producer but especially of the donor on his purchase and the time he spent searching for it, selecting and demonstrating affection in attention to the receiver's taste.

It is not too farfetched to associate the Maori sense of *hau* to what Walter Benjamin called the "*aura*" in an artwork (I now notice even casual phonetic similarity). (Benjamin 1968, 217– 251) His thesis on the loss of *aura* in art in the age of mechanical reproduction can also explain the loss of *hau*. Fetishization according to Marx results when erasing the work of the producer in the industrial manufacture of the object, namely, every trace of human *hau*, which prevents workers from recognizing themselves in the product of their labor and being appreciated as such by buyers. (Marx 1906, 81–96) Belief in the *hau* or *aura* has been exploited by advertising and marketing companies that now manufacture and simulate a pseudo-*hau*, a farce of personality for commodities. The elegance of Catherine Deneuve is sold in a bottle of Chanel and the agility of sports celebrities of the moment in a pair of sneakers.

In the artistic matrix, an additional dimension in Marcel Duchamp's conceptual work, particularly his *Fountain*, becomes significant from this perspective: it is an act of choosing an industrial, anonymous object, totally *aura*-less, and conferring upon it a kind of spirit or *hau* by signing it and declaring it a work of art. In this sense, the urinal chosen by Duchamp becomes unique and different from the thousands of identical others of the same series because only this one possesses the particular *hau* conferred upon it by the artist. For Duchamp, authorship became a matter of declaration rather than fabrication; a non-verbal performative speech act of nomination or conferral of the status of art upon an object by mere whim and much perspicuity. This excess in a mere signature, totally superfluous for performing the functional act in it, suffices to mutate it into an artwork.

Another case of Western contemporary production of *hau* is the so-called car art. Andy Warhol, Roy Lichtenstein, Robert Rauschenberg, and David Hockney, among others, each turn a BMW, an expensive piece of machinery, into an even more expensive (and far less useful) work of art. These vehicles must now be carefully packed and shipped to end up exhibited in total immobility at museums as art. Between a regular car and "car art," the difference is the *hau* of the artist who painted it. This spirit that demands reciprocation in archaic societies, in modernity justifies an exorbitant price without regard to the amount of work invested or any practical advantage derived from the object. The incredibly high price of $5.6 million that was paid for Marilyn Monroe's iconic dress (of the subway wind picture) at an auction is another corroboration of Western attraction for the *hau* of objects.[7]

LIMITS OF EXCESS AND THE LAW OF THE CYCLICAL DYNAMICS

All that Marcel Mauss can tell us regarding the aesthetic is just an allusion:

> [A]ll kinds of institutions are given expression at one and the same time—religious, juridical, and moral, which relate to both politics and the family; likewise economic ones, which suppose special forms of production and consumption, or rather of performing total services and of distribution. This is not to take into account the aesthetic phenomena to which these facts lead, and the contours of the phenomena that these institutions manifest. (Mauss 1990, 3)

What does Mauss mean by saying that these facts lead to aesthetic phenomena? I will venture a response. Mauss called "total social l phenomenon" the case cited above. The Paleolithic cave paintings integrated and expressed the power of man upon his reality, his magic to summon the animal for hunting, a communication strategy among hunters for planning and carrying out an attack, the elaboration of the best possible hunting instruments, and techniques, all operating as aesthetic and semiotic devices.[8] Similarly, the ceremonies of the dissipation of excess like the potlatch and mayordomía combine the several dimensions: economic, aesthetic, social, semiotic, legal, sacred, and political. Mauss and Malinowski therefore hoped to find in *kula* circulation and potlatch patterns of the origins of the economy, law, religion, and morality, or bridges between *secundus* and *tertius*.

These "total services of an agonistic type" are performed for the sake of producing *mana* or honor and prestige for the host. It is not clear, however, why or how destroying things of value—that have a soul or *hau*—can generate prestige or *mana*. If there is a power in things that compels reciprocity through gift exchange among archaic societies, why wouldn't this same power impel their preservation rather than their destruction? For instance, the power of time, of human labor, of information that force us to preserve historical documents seem to be contradicted by the practice of the potlatch. There must be another reason, in addition to *hau* or "spirit of things" or "soul of their maker" that can explain the process of reciprocity and help us to understand how the destruction of useful objects produces prestige.

In fact, Darwinism provides the answer to the question formulated by Mauss: "What rule of legality and self-interest, in societies of a backward or archaic type, compels the gift that has been received to be obligatorily reciprocated? What power resides in the object given that causes its recipient to pay it back?" (Mauss 1990, 3) Such power is the reciprocity/status coupling that we inherit at least from our primate ancestors, and is present as well at least in birds and bats. (Wright 2009, 52, 57)

But reciprocity in the *tertius* manifests an additional feature: if one should give almost everything one has to preserve one's place and balance in the community, it must at least be performed with style and character. Expenditure should be a memorable event in an aesthetically deployed ritual with dramatic force that makes a difference between mere destruction of property in nature and aesthetic ostentatious dissipation in culture. To destroy what one owns is also to display detachment from properties. In other words, the gift must be of an excess and in excess.

THE AESTHETIC DISPLAY OF EXCESS

According to Mauss, certain ceremonies have to be performed because "to make a gift of something to someone is to make a present of some part of oneself. . . . To retain that thing would be dangerous and mortal." (Mauss 1990, 12) The reason for compulsory reciprocity lies, therefore, less in the *hau* or spirit of the thing retained, than in the act of retaining it. At issue here is the attitude towards retaining or giving and differentiates Western retentive capitalist societies from expulsive communities like the Häida and Tlingit. This difference resides in a form of fluxion understood as centripetal or centrifugal disposition to open or close, tighten or relax, spend or contain, control or dissipate energy, matter, or time in the social, artificial, or natural environment.[9]

Certain cultures exhibit open centrifugal fluxion through proud traditions of publicly giving away in rituals such as the potlatch or the anniversary feast of the town or *barrio's* (neighborhood) patron saint. Such usages, as the Mexican tradition of *niñopan* in Xochimilco barrio, commit the mayordomo to cover the costs of dancers, musicians, candles, incense, fireworks, drinks and dinner, to keep the saint surrounded by flowers all year round and celebrate its day feeding the entire community. What's more, each guest takes home a package of food or *itacate* to feed her family for several days, a pre-Hispanic tradition still practiced in Iztapalapa, Xochimilco, and other districts and municipalities in Mexico.

Western capitalist economies, in contrast, maintain a centripetal flow and what they value is the power of accumulation, since prestige and honor are the result of hoarding this excess. They build monumental mansions, sometimes a number of them, to keep luxury items (silverware, china, fine furniture, art collections) and to display them in front of guests for admiration. Bank accounts are personal and private, but the display of excess by art collections is considered in good taste because of its aesthetic alibi, as if it could be possible to equally share the beauty of the object through contemplation alone and not lose any of it, on the contrary. Furs, jewelry, and automobiles make the cost visible. In the opulence of weddings and birthday parties or political campaigns, investment is calculated as a public exhibition of political, social, and pecuniary capital and thus increasing the symbolic assets of the message sender: All in exchange for status or prestige and nothing in return.

For centripetal cultures luxury consumption is indicative, as Veblen (1924) realized, not of how much the hosts are able to offer but of how much they still own to squander such amounts. Since accumulation for Western capitalist cultures can be unlimited, excess is literally impossible because there is always room for more. One is allowed to accumulate unscrupulously real estate, art collections, bank accounts, Blahnik shoes, designer suits, antiques, gadgets, yachts, sexual partners, sports cars, pets, and pedigree animals. There is one case, however, where excess and accumulation are strictly forbidden by current Western standards: body fat. Being overweight is seen as the antithesis of aesthetics severely punished by fashion. Celebrities and sex idols exhibit their control over fat, sometimes explicitly specified in their professional contracts. The new French cuisine consists of expensive dishes cooked for the visual more than the palate, served on heavy white square ceramic plates in tiny portions of slim line protein all very decorated and ready for contemplation. There is a real fear of excessive weight, but there is no law prohibiting obesity in material properties as those proudly exhibited in Forbes magazine.

This strange logic of prestige and aesthetics does not work in more traditional archaic communities. Whatever the weight of a person is his own personal matter and symbol of wealth, while the accumulation of objects is despised in potlatch-observing communities as much as the accumulation of body fat in the jet-set society. I find a prohibition to accumulate wealth much more reasonable than the taboo to accumulate fat, since wealth is social in character with a direct impact on the community, while fat is personal and a private matter. From this point of view, what distinguishes these cultures is not so much their fetishes or animistic superstitions regarding *hau*, but the contrast in boundaries each one establishes between the necessary and the excessive to get the status.

The non-Western version of Adam Smith's definition could then be written as follows: excessive is what the community considers indecent to accumulate. The *hau* in non-Western societies imposes reciprocation and demands circulation, whereas the *aura* in Western societies is purchasable, collectible, and increases exchange value. In sum, the logic underlying obligatory reciprocity would appear to depend less upon the *hau* of things observed by Mauss, than upon a dynamic and communal sense of life, of the world, of work, and of its products. As I mentioned above, it is an attitude towards retention rather than towards what is retained. This reciprocity considers as mere common sense that we should give when we receive, as in biological processes of inhaling and exhaling, birth and death, planting and harvesting, but which seems incomprehensible to the economic system of capital accumulation.

The circulation of matter and energy, the movement of the sea and the moon, the changing seasons show a dynamic of giving and taking, of abundance and dynamism, not of scarcity and immobility. The "total social phenomenon" goes beyond the social to encompass the cosmic since the destruction of wealth is its partial return to the gods.

Offering almost everything one owns in a potlatch ceremony or mayordomía is only possible if one understands the economy of excess as described by Bataille. From this perspective, it expresses a belief that what is lost will be recovered in one way or another by someone, a faith in the generosity of life. This view of the social, natural, and cosmic clarifies reciprocity rituals among the societies studied due to the belief that everything has a spirit that can take revenge and must be kept moving. For nomadic groups it is an elemental practical consideration giving away objects to get rid of them since wealth is a burden. Accumulating is, in this context, *contra natura*, equivalent to hijacking something destined to be in motion.

The difference of the *hau* (that obliges circulation) and the *aura* (demands accumulation) also expresses an opposite attitude to the relationship between individual and community: the *hau* integrates the individual to the group and the *aura* separates him from it. What is remarkable is the crossroad where such different cultures seem to converge: the transmutation of material goods into symbolic capital through dissipation rituals of excess via the aesthetic.

EXCEEDING

Kant's understanding of aesthetic appreciation as "pure disinterested delight" implies that it does not produce any kind of benefits, does not generate concepts, and is an end in itself. This sense of gratuitousness seems to be Bataille's indirect debt to Kant when he argues that Aztec human sacrifice pretended to set humankind beyond the utilitarian, liberating sacrificial victims from servility. Although Bataille (1987) was wrong in attributing to the Aztecs the belief that sacrifice was gratuitous (as he neglected mythical signification in which they regarded sacrifice not only as useful but indispensable for the survival of the sun-god Huitzilopochtli and of the entire community), he nonetheless was accurate concerning the human need to transcend the utilitarian and feel free from restrictions. Had Aztec human sacrifice indeed been, as Bataille argued, non-utilitarian, it would have been even more monstrous, hardly discernible from killing for the sake of killing. Available evidence in the context of the mythological meaning they attributed to this act shows that sacrifice was almost a technological device for preserving the world according to Aztec religion. In their world view, it was simply logical that the sun-god, like humans, needed to be fed. Human sacrifice for the Aztecs was both logically and socially necessary, not excessive.

The way the Aztecs did exceed the utilitarian resided not only in the number of victims for sacrifice to the sun-god, but in the style in which it was performed, in the splendor of their temples and pyramids, in their spectacular rituals, festivals, and celebrations, in the sumptuous lifestyle of their *tlatoani* or leader, and in the tzomplantli poles exhibiting as much as 60,000 skulls of sacrificed victims in Tenochtitlan. These customs structured the aesthetics of the Aztec military and theocratic regime. The excess was not in the sacrifice itself but in its aestheticization. By aestheticizing rituals, the *tlatoani* and the priesthood vicariously shared them with the people. This technology followed the law of obligatory reciprocity in both directions: the sun god Huitzilopochtli had to reciprocate human sacrifice by ensuring the survival of the community, and the community had to return via the sacrificed the favor of the god's daily battle against darkness. Every sacrifice obliges, from monkeys' grooming and bats' regurgitation to the crucifixion of the Christ for the sins of devotees.

Adolfo Sánchez Vázquez, following Kant, also noted that the aesthetic occurs as an excess that goes beyond strict instrumentality and generates a function that opens up a new space into the aesthetic.

> It is difficult to believe that prehistoric man wouldn't have certain awareness that, in introducing certain *excedent* formal elements he was *exceeding* the practical utilitarian limits beyond which a new space would be opened: precisely what we call the aesthetic. . . . Furthermore, in producing it—not by chance—certain formal elements that *exceeded* the functional requirements, there also had to be some awareness of one's ability to create them. (Sánchez Vázquez 1992, 100; emphasis added.)

Sánchez Vázquez, however, diverts from Kant's conception, as he does not say that the aesthetic is non-utilitarian but that it is meta-utilitarian. In other words, we have here a function that is still practical not as a physical utensil but as a symbolic item. Decorated flint tools, for example, display gratuitousness by exceeding their strictly instrumental shape, which in turn produces a shift from the original function involved. Functionality is not annulled, as these tools are still useful in a different sense: "The product of labor, by putting to work the capacity to transform matter, continues being an instrument, albeit a symbolic one, that enables it—thanks to its form—to act upon the real." How does this shift from instrumental to symbolic functionality occur? Sánchez Vázquez unknowingly agrees with Jakobson's concept of the aesthetic as consisting in the form itself when he states that "a product fulfills this symbolic function when it acquires a "good form" (or "excessive form") as a result of "good work." He adds:

> All the creative power of painter-hunters of the Upper Paleolithic period, culminating what was accomplished during thousands and thousands of years of work—and among its highest achievements was the capacity to endow the material with an "excessive" form—was symbolically, magically put to the service of the practical end of hunting wild animals. (Sánchez Vázquez 1992, 100–1)

Returning to the original question—why excess produced in a social group is dissipated instead of being fairly distributed—the answer seems to involve the antipode of excess or the realm of utmost precariousness: mortality. Material excess may seem limitless, but an individual's life is fatally limited. Consequently, whenever there is material excess, it holds the secret wish to be exchanged for wealth in additional time. How can wealth in time be attained? This can be achieved symbolically through a transition from the personal to social realm by vicarious permanence in the memory of other members of the community. The aesthetization of material excess preserves the *hau* or spirit of things by the mediation of the aesthetic, which recirculates it and converts material dissipation into symbolic accumulation of prestige or mana conveyed by fellow men. Western art was interpreted by Freud as a process of sublimation, a permutation of libidinal energy into social, cultural, and symbolic wealth. Material wealth is private or personal, whereas symbolic wealth (as language, prestige, or power) is always social and conferred by others. The pleasure by aesthetization of excess can be shared with the gods and the community through rituals such as potlatch, mayordomía, carnivals, sacrifices, or other religious ceremonies. There is no culture that does not put special efforts into the beautification of the objects of worship.[10]

The belief that the gods are aesthetically sensitive and that, to appeal to divine benevolence, offerings should be aestheticized if we pretend to capture their attention, appears to be universal. Dewey emphasizes that "music and singing were intimate parts of the rites and ceremonies in which the meaning of group life is consummated." (Dewey 1980, 6–7) We have here various permutations that echo each other: material to symbolic, private to public, corporeal to mental, ephemeral to durable, individual to communal, mortal to immortal, subjective to objective. All these are strangely mediated by aesthetics expressing the former by the latter as the genotype is expressed by the phenotype.

When there is trust in the generation of excedents, a significant part of the population is engaged in aesthetic activities at the service of the state, religion, and art. Material excess invested in pyramids and towering temples does not attenuate the material needs of the community, but it does exhibit a local heritage which embodies the pride of the polis, such as the Parthenon. The higher the magnificence of the monuments, the greater the confidence the community could have in their ability to survive and at the same time overpower the individual by the massive architectural scale.

While Aristotle advised temperance, what we really enjoy is excess: It assures us that life is magnanimous and the world exuberant. Hence, in a context that is bountiful, it is only natural to be generous. It has, in this sense, ethical repercussions. Strict calculation and control over people's time, desires, and energy, as occurs in totalitarian, capitalist, and bureaucratic regimes, stems from a sense of a precarious reality. By hoarding it leads, as Bataille insisted, to war and uncontrollable destruction. If, as Bataille argues, excess is logically necessary in economy as in nature, then aesthetic, as its symbolic transfiguration and gratifying way of dissipating it, is functionally and socially necessary.

We need to trust the reality of excess, to feel that we can lose without remorse, that there is a margin for vagary and play, that life gives more than we can take. This necessary confidence is nowhere better conveyed and expressed than by aesthetic concretization. If excess is inevitable and its dissipation imperative, then the aesthetic is indispensable.

NOTES

I wish to thank the publishers for granting permission to reprint part of the following article: Mandoki, Katya, "Material excess and aesthetic transmutation," in *Economies of Excess: Parallax*. John Armitage (ed.). Issue 18 Vol. 7 No. 1. (Jan–March 2001). pp. 64–75. London: Taylor & Francis. ISSN 1353–4645 DOI:10.1080/13534640010015935. A previous version of this text was published in Mandoki, Katya, "The Indispensable Excess of the Aesthetic." *Filozofski Vestnik–Acta Philosophica* 20, no. 2 (1999): 173–79.

1. Explained below in section The Aesthetic Display of Excess.
2. In the narrow sense of aesthetics as art, Weiss proposes to explore excess (sexual, libidinal, and emotional) in relation to madness, perversion, and psychoanalysis through Sade, Breton, Artaud, Bataille, Nietzsche, Klossowski. (Weiss 1989)
3. For a discussion and criticism of this aspect in Lacanian psychoanalysis see Sheets-Johnstone (1994, 311–17).
4. On signic and symbolic orders, see Mandoki (1994, 2007).
5. Moreno quoted in Zahavi and Zahavi (1997, 173).

6. "Exuberance is Beauty," in "Proverbs of Hell," *William Blake Complete Works Ultimate Collection: 250+ Works All Poetry, Poems, Prose, Annotations, Letters, Rarities Plus Biography* [Kindle Edition]. Everlasting Flames Publishing, 2013.

7. Duke, Alan. "Marilyn Monro's Iconic Dress Sells for $5.6 Million at Auction," *CNN.com*. http://www.cnn.com/2011/SHOWBIZ/celebrity.news.gossip/06/19/hollywood.auction/. Accessed 4 July 2014.

8. Cf. Heyd (2005).

9. On the concept of fluxion (Mandoki 2007, 152–54).

10. See on this aspect Dissanayake's (1990, chapter 4) "Making Special."

Epilogue

We have arrived at the end of this inquiry into the aesthetic dimension from the triple evolutionary, biosemiotic, and pragmatist perspective. Our starting point was aesthesis at *zero degree* in inert materiality considering that the exact opposite of the aesthetic is not ugliness (as commonly thought) or logic (as implied in Kant) but the insensitivity of inanimate matter. The basic condition for aesthesis to emerge is the concrete materiality of this world, *Orbis primus*, not the insubstantial spirituality of any other.

From there we made an infinite leap into *degree one* of aesthesis that took place at least 3.7 giga years ago when matter organized itself into replicating patterns. As replication requires nourishment, matter opened up sensing itself and its surroundings. Hence the miraculous materialization of aesthesis occurred at that precise instant when someone sensed something somewhere. Organisms evolved in their tenacious orientation to light, moisture, gravity, and nutrients thus populating the *Orbis secundus*. Regardless of how simple a creature is, we can say that some sense of joy is present when satisfying needs to survive (quench its thirst, flee a predator, copulate), and a sense of pain when wounded, hungry, locked up, or threatened. Emotionality and aesthesis are woven into life itself.

Degree two of aesthesis emerged when organisms were able not only to detect environmental stimuli but to communicate and act upon each other to provoke a reaction. The emission of molecular messages to repel predators, as acacias discharging tannin to alert others, or various species releasing pheromones to signal alarm or to attract, to aggregate, and to mark territories, all result from this immense jump from one threshold to another that enables interaction and collective organization.

It took yet another evolutionary leap to reach the *third degree* of aesthesis, or cultural sensibility, expressed at the precise instant when a hominid freezes time 2.5 million years ago by inventing the first artifact projected to the future. He carves bi-facial axes from stones and marks an occasion, an idea, or a quantity on a bone. She models a Venus in clay, leaves a trace of her hand on the rock, paints images in caves and parchments, agrees on a sound to denote a thing, makes tools and decorates them, colonizing the *Orbis tertius* with mountains, rivers, and valleys of artifacts.

Sensibility, as the condition for perception, appreciation, and creation, thus unravels the orbicular enigma of aesthetics: does beauty mark a direction to evolution, or is it that such direction *is* what we call beauty? In a way, Kant intuited this direction when he wrote that the beautiful "directly brings with it a feeling of the furtherance of life." (Critique of Judgement § 23) As we observe the efforts of the paradise bird in New Guinea, so excessively endowed with charm in its plumage, its song, its antics, and its dance for attracting the female to impel life forward through offspring, we witness the gratuitous utility of beauty at work. Was it necessary? Is it superfluous? It is certainly excessive, yet absolutely indispensable. Aesthetics are, evidently, the key to selection and to directionality in evolution.

As a colophon, I now realize *that obscure object of desire* is elusive and *the sublime object of ideology* just kitsch. *The unbearable lightness of being* blows us away while *the indispensable excess of the aesthetic* lures evolution to exceed herself in beauty.

We admire magnificent designs in birds, fish, flowers, and beetles because we are ourselves a bit flower, bird, fish, and beetle. Hence, it suddenly becomes so clear that Kant's "transcendental aesthetics" are actually nothing other than the aesthesis of all those creatures who preceded us and whose genes and neurons are entwined with our own anatomy and perception. Each mutation a live organism underwent in its resolute urge to be fruitful and multiply, and each choice a female made for mating guided by beauty, have bestowed upon us a world so generous in its variety and splendor that it must be qualified as sacred. By the time we arrived as humans onto the scene, aesthesis had long been thrusting nature to avidly explore the world, illuminate it through every creature and enjoy.

תושלב"ע

Glossary

Abstimmung: term used by Parret referring to tuning of organisms with their environment.

Aesthesis: sensitivity, perception, receptivity.

Aesthetics: study of aesthesis.

Aura: term used by Walter Benjamin to denote what characterizes the particular attraction to artworks and connotations due to their authorship, singularity, skill, or antiquity.

Bio-aesthetics: theory that studies processes of aesthesis along the entire scope of creatures and species.

Biosemiotics: area of semiotics that specializes in the study of processes of semiosis in all living organisms.

***Cogitans* network**: cultural products as artifacts, languages, customs, memes.

Cyto-aesthetics: theory that studies the processes of perception and sensitivity at the cellular level.

Display: exhibiting oneself for a re-actor to be perceived in certain terms.

Effector: term coined by von Uexküll to denote the active part in the receptor–effector cycle in the life process of an organism.

Entropy: term from the theory of thermodynamics that refers to the tendency towards disorder and homogeneity in a closed system.

Evolution: paradigm that explains the process of transformation of the species in nature by the mechanism of random mutation and retentive selection.

***Extensa* network**: term taken from Descartes to refer to the expanding physical world, e.g., quarks, photons, radiation, atoms, planets, electromagnetism, and so on.

Firstness, secondness, thirdness: operational concepts of the phaneroscopy or phenomenology in Peirce.

Fitness: for evolutionary theory equals survivability and reproduction.

Genocentrism: evolutionary theory that explains all human behavior based on the mechanism of genetic replication. (Dawkins)

Genotype: organization of genetic information in the individual.

Handicap principle: exhibition of a disadvantage to credit honesty of signals according to Zahavi.

Hau: term used by Mauss to refer to the spirit of things, which can retain part of the soul of its maker.

Hylomorphism: from *hylos* (matter) and *morphe* (form), terminology used by Aristotle.

Natural selection: the mechanism by which the evolutionary paradigm explains the process of evolution.

Negentropy: negative entropy or tendency to order. (Schrödinger)

Ontogenesis: evolutionary process of an individual throughout his whole life.

Ontopoietics: deliberate elaboration or preparation of an act or artifact.

Orbis primus: "first world"—inanimate, physical reality.

Orbis secundus: "second world"—the emergence of life, or the biosphere.

Orbis tertius: "third world"—the cultural world, or artifacts.

Perceptor: term to denote the function of reception by an agent in the functional cycle that complements the effector. (von Uexküll)

Phenotype: macroscopic external manifestation of the genotype.

Phylo-genesis: evolutionary history of a species over generations.

Phylo-poietics: indirect conformation of the species over generations by sexual selection.

Phyto-aesthetics: theory that studies the processes of perception and sensitivity at the level of plants and vegetation.

Poetics: term used by Aristotle to denote the study of the aesthetics of art, particularly in the case of Tragedy.

Poiesis: the process of elaborating or creating an act or artifact.

Potlatch: custom among certain ethnic groups to publicly exhibit the squandering of goods.

Praxis: daily labor of survival.

Presentamen (plural: presentamina): phenomenon or stimulus presented to the perception of the interpretant. (Peirce)

Prosaics: the study of aesthetics in everyday life. (Mandoki)

Quorum sensing: perception of critical mass or proximity of other bacteria in the *Vibrio fischeri*.

Random variation and **natural selection**: formula for the mechanism of evolution, whereby random changes in genetic information are selected when they are beneficial to the fitness of the organism.

Re-actor: destinatary of a sign to whom an actor directs a message to achieve a particular reaction.

Semiosis: processes of signification and interpretation studied by semiotic theory.

Signaling: information displayed by an animal to communicate its status.

Significans **network**: signifying flow by live creatures from unicellular to multicellular species. (Mandoki)

SPN: singular proper name. (Sebeok)

Symbiogenesis: merging of two organisms in symbiosis to form one. (Margolis)

Umwelt: bubble that encompasses the individual of different species that determines perception according to their anatomy.

Zoo-aesthetics: theory that studies the processes of perception, expression, and sensitivity of different animal species. (von Uexküll)

Bibliography

Abbagnano, Nicola. *Diccionario de Filosofía*. México, DF: Fondo de Cultura Económica, 1963.

Anderson, Myrdene. "Rothschild's Ouroborus." *Sign Systems Studies* 31 (2003): 301–14

Appleton, Jay. "Prospects and Refuges Re-Visited." *Landscape Journal* 3, no. 2 (1984): 91–103.

Arak, Anthony, and D. Magnus Enquist. "Conflict, Receiver Bias and the Evolution of Signal Form." *Philosophical Transactions: Biological Sciences* 349, no. 1330 (1995): 337–44.

Arendt, Hannah. *The Human Condition*. Chicago: Univ. of Chicago Press, 1998.

Aristotle. "De Anima." *Aristotle: De Anima Electronic Text Center*, University of Virginia Library, n.d. Accessed February 22, 2013. http://etext.lib.virginia.edu/toc/modeng/public/AriSoul.html.

———. *De sensu et sensibili.* Electronic Text Center, University of Virginia Library. http://etext.virginia.edu/toc/modeng/public/AriSens.html. Accessed February 22, 2013.

Armstrong, Susan Jean, and Richard George Botzler. *The Animal Ethics Reader*. New York: Routledge, 2008.

Aunger, Robert. *The Electric Meme: A New Theory of How We Think*. New York: The Free Press, 2002.

Aviezer, Nathan. "The Anthropic Principle: What Is It and Why Is It Meaningful to the Believing Jew?" *Jewish Action* Spring 575 (1999). http://www.ou.org/publications/ja/5759spring/anthropic.pdf.

Avital, Eytan, and Eva Jablonka. *Animal Traditions: Behavioural Inheritance in Evolution*. Cambridge: Cambridge University Press, 2000.

Aydon, Cyril. *Charles Darwin: The Naturalist Who Started a Scientific Revolution*. New York: Carroll & Graf, 2002.

Bahn, Paul G., and Paul Pettit. "Ice Age Art: Arrival of the Modern Mind." *Antiquity Publications* 15, no. 7 (2013): 8.

Baldwin, Mark. "Organic Selection." *Science*, New Series 5, no. 121 (1897): 634–36.

Barber, Nigel. "Mustache Fashion Covaries with a Good Marriage Market for Women." *Journal of Nonverbal Behavior*. Vol. 25. Springer Netherlands, 2001. doi:10.1023/A:1012515505895.

Barbieri, Marcello. *Introduction to Biosemiotics: The New Biological Synthesis*. New York: Springer, 2008a.

———. "The Code Model of Semiosis: The First Steps Toward a Scientific Biosemiotics." *The American Journal of Semiotics* 24, no. 1–3 (2008b): 23–37.

———. "The Ribotype Theory on the Origin of Life." *Journal of Theoretical Biology* 91, no. 4 (1981): 545–601.

Barkow, Jerome H., Leda Cosmides, and John Tooby. *The Adapted Mind: Evolutionary Psychology and the Generation of Culture*. Oxford : Oxford University Press US, 1992.

Bartra, Roger. *Antropología Del Cerebro: La Conciencia Y Los Sistemas Simbólicos*. Mexico: Fondo de Cultura Economica, 2007.

Bataille, Georges. *La Parte Maldita*. Barcelona: Icaria Editorial, 1987.

Bateson, Gregory. *Steps to an Ecology of Mind*. Chicago: University of Chicago Press, 2000.

Bateson, Gregory, and Rodney E. Donaldson. *A Sacred Unity: Further Steps to an Ecology of Mind*. Glenview: Harper Collins, 1991.

Baumgarten, Alexander Gottlieb. *Aesthetica Scripsit Alexand. Gottlieb Bavmgarten . . .* Hildesheum, Zurich, New York: Georg Olms Verlag, 1750. https://play.google.com/books/reader?id=iPVNLFAGznQC&printsec=frontcover&output=reader&hl=en&pg=GBS.PA3.

Bednarik, Robert G. "A Figurine from the African Acheulian." *Current Anthropology* 44, no. 3 (June 20, 2003): 405–13. Accessed July 8, 2014. doi:10.1086/374900.

Bekoff, Marc, and John Alexander Byers. *Animal Play: Evolutionary, Comparative, and Ecological Perspectives*. Cambridge: Cambridge University Press, 1998.

Ben Shlomo, Yosef. *Poetry of Being; Lectures on the Philosophy of Rabbi Kook*. Tel Aviv: Mod Books, 1990.

Benjamin, Walter. "The Work of Art in the Age of Mechanical Reproduction." In *Illumination*, edited by Walter Benjamin, 217–51. New York: Hartcourt, 1968.

Berger, Peter L., and Thomas Luckmann. *The Social Construction of Reality: A Treatise in the Sociology of Knowledge*. New York: Doubleday, 1966.

Berleant, Arnold. *Sensibility and Sense: The Aesthetic Transformation of the Human World*. Charlottesville: Imprint Academic, 2010.

Bertalanffy, Ludwig Von. *General System Theory. Foundations, Development, Applications*. New York: George Braziller Inc., 1969.

Biggers, J. D. and R. F. S. Creed. "Two Morphological Types of Placentae in the Raccoon." *Nature* 194 (1962): 103–5.

Bilko, A., V. Altbacker, and R. Hudson. "Transmission of Food Preference in the Rabbit: The Means of Information Transfer." *Physiology & Behavior* 56, no. 5 (November 1994): 907–12. doi:10.1016/0031-9384(94)90322-0.

Binkley, Timothy. "Piece: Contra Aesthetics." *The Journal of Aesthetics and Art Criticism* 35, no. 3 (Spring, 1977): 265–77.

Blackmore, Susan. "Evolution and Memes: The Human Brain as a Selective Imitation Device." *Cybernetics and Systems: An International Journal, 1087–6553* 32, no. 1 (2001): 225–55.

———. *The Meme Machine*. Oxford University Press, 2000.

Bloch, Maurice. "Language, Anthropology and Cognitive Science." *Man, New Series* 26, no. 2 (1991): 183–98.

Borges, Jorge Luis. *Tlön, Uqbar, Orbis Tertius*. México D.F.: Siglo Veintiuno Editores, 1968.

Borgia, Gerald. "Bower Quality, Number of Decorations and Mating Success of Male Satin Bowerbirds (Ptilonorhynchus Violaceus): An Experimental Analysis." *Animal Behaviour* 33, no. 1 (February 1985): 266–71. doi:10.1016/S0003-3472(85)80140-8.

———. "Complex Male Display and Female Choice in the Spotted Bowerbird: Specialized Functions for Different Bower Decoration." *The Association for the Study of Animal Behavior*, 1995, 1291–1301.

———. "Satin Bowerbird Parasites: A Test of the Bright Male Hypothesis." *Behavioral Ecology and Sociobiology* 19, no. 5 (November 1986): 355–58. doi:10.1007/BF00295708.

Borgia, Gerald, and Ken Collis. "Female Choice for Parasite-Free Male Satin Bowerbirds and the Evolution of Bright Male Plumage." *Behavioral Ecology and Sociobiology* 25, no. 6 (December 1989): 445–53. doi:10.1007/BF00300191.

Borgia, Gerald, and J. Keagy. "An Inverse Relationship between Decoration and Food Colour Preferences in Satin Bowerbirds Does Not Support the Sensory Drive Hypothesis." *Animal Behaviour* 72, no. 5 (November 2006): 1125–33. doi:10.1016/j.anbehav.2006.03.015.

Borgia, Gerald, and Mauvis Gore. "Feather Stealing in the Satin Bowerbird." *Animal Behavior* 34 (1996): 737–38.

Bostwick, Kimberly S., and Richard O. Prum. "High-Speed Video Analysis of Wing-Snapping in Two Manakin Clades (Pipridae: Aves)." *Journal of Experimental Biology* 206, no. 20 (2003): 3693–706.

Boyd, Robert, and Peter J. Richerson. *Culture and the Evolutionary Process*. The University of Chicago Press, 1985.

———. *Culture and the Evolutionary Process*. University of Chicago Press, 1988.

———. *The Origin and Evolution of Cultures*. Oxford University Press US, 2005.

Brainard, M. S., and A. J. Doupe. "Auditory Feedback in Learning and Maintenance of Vocal Behaviour." *Nature Reviews. Neuroscience* 1, no. 1 (October 2000): 31–40. doi:10.1038/35036205.

Brandt, Per Aage. "Form and Meaning in Art." In *The Artful Mind: Cognitive Science and the Riddle of Human Creativity*, edited by Mark Turner. Oxford University Press US, 2006.

Breidbach, Olaf. "The Beauties and The Beautiful—Some Considerations from the Perspective of Neuronal Aesthetics." In *Evolutionary Aesthetics*, edited by Karl Grammer and Eckart Voland, 39–68. New York: Springer, 2003.

Brighton, Henry, Simon Kirby, and Kenny Smith. "Cultural Selection for Learnability: Three Principles Underlying the View That Language Adapts to Be Learnable." In *Editor, Language Origins: Perspectives on Evolution*, edited by M. Tallerman. Oxford: Oxford University Press, 2005.

Brockman, John, ed. *The Third Culture: Beyond the Scientific Revolution*. New York: Touchstone Books, 1996.

Brum, P. E. D., L. E. Costa-Schmidt, and A. M. d. Araujo. "It Is a Matter of Taste: Chemical Signals Mediate Nuptial Gift Acceptance in a Neotropical Spider." *Behavioral Ecology* 23, no. 2 (December 08, 2011): 442–47. doi:10.1093/beheco/arr209.

Bruni, Luis Emilio. "Cellular Semiotics And Signal Transduction." In *Introduction to Biosemiotics: The New Biological Synthesis*, edited by Marcello Barbieri, 365–408. Dordrecht: Springer, 2008.

Brusca, Richard C., and Gary J. Brusca. *Invertebrados*. Madrid: McGraw Hill, 2005.

Burghardt, Gordon M. "The Evolutionary Origins of Play Revisited: Lessons from Turtles." In *Animal Play: Evolutionary, Comparative, and Ecological Perspectives*, edited by Marc Bekoff and John Alexander Byers, 1–23. Cambridge: Cambridge University Press, 1998.

Burki, Fabien, and Et Al Shalchian-Tabrizi, Kamran. "Phylogenomics Reshuffles the Eukaryotic Supergroups." *PLoS ONE* 2, no. 8 (2007).

Butvill, Dave Brian. "Wild Lives: Mimic Octopus — the Shape-Shifter." *California Wild* 56, no. 3 (2003): 38–40.

Caillois, Roger. *Los Juegos Y Los Hombres; La Máscara Y El Vértigo*. México D.F.: Fondo de Cultura Económica, 1994.

Canetti, Elias. *Masa Y Poder*. Barcelona: Alianza Editorial, 2005.

Capra, Fritjof. *La Trama de La Vida: Una Nueva Perspectiva de Los Sistemas Vivos*. Barcelon: Anagrama Editorial S. A., 2002.

Carter, Brandon. "Large Number Coincidences and the Anthropic Principle in Cosmology." In *Confrontation of Cosmological Theories with Observation Data, Proceedings of the Symposium, Krakow, Poland, September 10–12, 1973*. Dordrecht: Reidel Publishing, 1974, 291–98.

Cavalli-Sforza, Luigi Luca, Marcus W. Feldman, and K. H. Chen. "Theory and Observation in Cultural Transmission." *Science* 218, no. 4567 (1982): 19-27. DOI: 10.1126/science.7123211.

Cavalli-Sforza, Luigi Luca, and Marcus W. Feldman. *Cultural Transmission and Evolution: A Quantitative Approach*. Princeton: Princeton University Press, 1981.

Churchland, Paul. "Eliminative Materialism and the Propositional Attitudes." *Journal of Philosophy*, 78, no. 2 (1981): 67–90.

Cockburn, Cynthia and Ann Oakley. "The culture of masculinity costs all too much to ignore." *Guardian*, November 25, 2011. Accessed November 27, 2011.

Conard, Nicholas J. "A Female Figurine from the Basal Aurignacian of Hohle Fels Cave in Southwestern Germany." *Nature* 459, no. 7244 (May 2009): 248–52. doi:10.1038/nature07995.

Cosmides, Leda, and John Tooby. "Cognitive Adaptations for Social Exchange." In *The Adapted Mind*, edited by Jerome H. Barkow, Leda Cosmides, and John Tooby, 163–228. Oxford; New York: Oxford University Press, 1992.

Coss, Richard G. "The Role of Evolved Perceptual Biases in Art and Design." In *Evolutionary Aesthetics*, edited by Eckart Voland and Karl Grammer, 69–130. New York: Springer, 2003.

Cronin, Helena. *The Ant and the Peacock: Altruism and Sexual Selection from Darwin to Today*. Cambridge: Cambridge University Press, 1993.

Dabbs, James M., Denise de la Rue, and Paula M. Williams. "Testosterone and Occupational Choice: Actors, Ministers, and Other Men." *Journal of Personality and Social Psychology*, 59, no. 6 (1990): 1261–65.

Daly, Martin, and Margo Wilson. *Homicide*. New Jersey: Transaction Publishers, 1983.

Damasio, Antonio R. *Descartes' Error: Emotion, Reason, and the Human Brain*. New York: G.P. Putnam, 1994.

Danto, Arthur. "The Artworld." *The Journal of Philosophy* 61, no. 19 (1964): 571–84.

———. *The Transfiguration of the Commonplace*. Cambridge: Harvard University Press, 1981.

Darwin, Charles.. "On the Origin of Species by Means of Natural Selection, or the Preservation of Favoured Races in the Struggle for Life." *Darwin Online*, 1859. Accessed July 14, 2014. http://darwin-online.org.uk/content/frameset?itemID=F373&viewtype=text&pageseq=1.

———. "On the Movements and Habits of Climbing Plants." *Journal of the Linnean Society of London (Botany)* 9 (1865): 1–118. Accessed July 14, 2014. http://darwin-online.org.uk/converted/published/1865_plants_F834a.html.

———. *The Descent of Man, and Selection in Relation to Sex, Volume 2*. London: John Murray, 1871. Accessed July 14, 2014. http://darwin-online.org.uk/EditorialIntroductions/Freeman_TheDescentofMan.html.

———. *Origin of Species*. London: John Murray, 1876.

———. *Descent of Man, and Selection in Relation to Sex: Darwin Online Second Edition, Revised And Augmented.* John Murray, 1882. Accessed July 14, 2014. http://darwin-online.org.uk/content/frameset?viewtype=side&itemID=F955&pageseq=1.

———. *The Life and Letters of Charles Darwin, Including an Autobiographical Chapter, Volume 2*. Edited by Francis Darwin. London: John Murray, 1887. Accessed July 14, 2014. http://darwin-online.org.uk/content/frameset?viewtype=text&itemID=F1452.1&pageseq=1.

Davies, Paul. "Our Place in the Universe." In *Modern Cosmology & Philosophy*, edited by John Leslie. New York: Prometheus Books, 1998.

Dawkins, Richard. *The Extended Phenotype: The Long Reach of the Gene*. New York: Oxford University Press, 1999.

———. *The Selfish Gene*. New York: Oxford University Press, 1976.

———. *Unweaving the Rainbow: Science, Delusion and the Appetite for Wonder*. Boston: Houghton Mifflin Harcourt, 2000.

De Waal, Frans B. M. "Bonobo Sex and Society: The Behavior of a Close Relative Challenges Assumptions about Male Supremacy in Human Evolution." *Scientific American* 272, no. 3 (1995): 82–88.

———. *Chimpanzee Politics: Power and Sex among Apes*. JHU Press, 2007.

Deacon, Terrence. "The Aesthetic Faculty." In *The Artful Mind: Cognitive Science and the Riddle of Human Creativity*, edited by Mark Turner, 21–55. Oxford; New York: Oxford University Press, 2006.

Deleuze, Gilles, and Félix Guattari. *Rizoma: Introducción*. Mexico D.F.: Coyoacan, 1966.

D'Errico, Francesco, and April Nowell. "A New Look at the Berekhat Ram Figurine: Implications for the Origins of Symbolism." *Cambridge Archaeological Journal* 10 (2000): 123–67. doi:10.1017/S0959774300000056.

Dewey, John. *Art as Experience*. New York: Perigee Books, 1980.

Diamond, Jared. *Guns, Germs and Steel*. New York London: Norton and Co, 1999.

Dickie, George. *Art and the Aesthetic: An Institutional Analysis*. New York: Cornell University Press, 1974.

Dinstein, Ilan, Cibu Thomas, Marlene Behrmann, and David J Heeger. "A Mirror up to Nature." *Current Biology : CB* 18, no. 1 (2008): R13–8.

Dissanayake, Ellen. "Aesthetic Experience and Human Evolution." *The Journal of Aesthetics and Art Criticism* 41, No. 2 (1982): 145–55.
———. *What is Art For?* Washington: University of Washington Press, 1990.
———. *Homo Aestheticus: Where Art Comes from and Why*. Washington: University of Washington Press, 1996.
———. "Komar and Melamid Discover Pleistocene Taste." *Philosophy and Literature*, 1998.
———. "What Art Is and What It Does: An Overview of Contemporary Evolutionary Hypotheses." In *Evolutionary and Neurocognitive Approaches to Aesthetics, Creativity and the Arts*, 1st ed. Amityville, New York: Baywood Publishing, 2007.
Dobzhansky, Theodosius. "Anthropology and the Natural Sciences—The Problem of Human Evolution." *Current Anthropology, Source* 4, no. 2 (1963): 138–48.
Donald, Merlin. "Art and Cognitive Evolution." In *The Artful Mind: Cognitive Science and the Riddle of Human Creativity*, edited by Mark Turner, 3–20. Oxford; New York: Oxford University Press, 2006.
Dondis, D. A. *La Sintaxis de La Imagen: Introducción Ai Alfabeto Visual.* Barcelona: Ed. Gustavo Gilli, 1976.
Dugatkin, L. A. "Interface between Culturally Based Preferences and Genetic Preferences: Female Mate Choice in Poecilia Reticulata." *Proceedings of the National Academy of Sciences* 93, no. 7 (April 1996): 2770–73.
Duke, Alan. "Marilyn Monro's Iconic Dress Sells for $5.6 Million at Auction," *CNN.com*. http://www.cnn.com/2011/SHOWBIZ/celebrity.news.gossip/06/19/hollywood.auction/.
Dunbar, Robin, Ian MacDonald, Chris Knight, and Camilla Power. *The Evolution of Culture: An Interdisciplinary View*. Edinburgh: Edinburgh University Press, 1999.
Dutton, Denis. *The Art Instinct: Beauty, Pleasure and Human Evolution*. New York: Bloomsbury, 2009.
Eagleton, Terry. *The Ideology of the Aesthetic*. New York: Wiley-Blackwell, 1990.
Eco, Umberto. *Apocalípticos E Integrados*. Barcelona: Debolsillo, 2004.
———. *Tratado de Semiótica General*. Barcelona: Nueva Imagen Lumen, 1976.
Edelman, Gerald M. *Bright Air, Brilliant Fire: On the Matter of the Mind.* New York: Basic Books, 1992.
Ekman, Paul. *Emotions Revealed: Recognizing Faces and Feelings to Improve Communication and Emotional Life*. Times Books, 2003.
Elias, Norbert. *The Civilizing Process: Sociogenetic and Psychogenetic Investigations*. Edited by Eric Dunning, Johan Goudsblom, and Stephen Mennell. New York: Wiley-Blackwell, 2000.
Endler, J. "Sensory Ecology, Receiver Biases and Sexual Selection." *Trends in Ecology & Evolution* 13, no. 10 (October 1998): 415–20.
Etcoff, Nancy. *Survival of the Prettiest: The Science of Beauty.* New York: Anchor Books, 2000.
Falk, Dean. *Finding Our Tongues: Mothers, Infants, and the Origins of Language*. New York: Basic Books, 2010.
Fernandez, Raquel, and Dani Rodrik. "Resistance to Reform: Status Quo Bias in the Presence of Individual-Specific Uncertainty." *The American Economic Review* 81, no. 5 (1991): 1146–55.
Fisher, Helen. *Anatomy of Love*. New York: Random House, 1992.
Fisher, Sir Ronald Aylmer. *The Genetical Theory of Natural Selection: A Complete Variorum Edition*. Edited by J. H. Bennett. Oxford: Oxford University Press, 1999.
Fodor, Jerry A. *The Modularity of Mind: An Essay on Faculty Psychology*. Massachusetts: MIT Press, 1983.
Fog, Agner. *Cultural Selection*. Dordrecht: Springer, 1999.
Foucault, Michel. *Las Palabras Y Las Cosas: Una Arqueologia de Las Ciencias Humanas*. Mexico D.F: Siglo Veintiuno Editores, 1986.
Galef, Bennett G. "The Question of Animal Culture." *Human Nature* 3, no. 2 (June 1992): 157–78. doi:10.1007/BF02692251.

Gallese, Vittorio, Luciano Fadiga, Leonardo Fogassi, and Giacomo Rizzolatti. "Action Recognition in the Premotor Cortex." *Brain* 119, no. 2 (April 1996): 593–609. doi:10.1093/brain/119.2.593.

Gardner, R. A., and B. T. Gardner. "Teaching Sign Language to a Chimpanzee." *Science* 165 (1969): 664–72.

Glaser, Edmund M. "Is There a Neurobiology of Hate?" *Journal Of Hate Studies* 77 (2009). http://guweb2.gonzaga.edu/againsthate/Journal7/Glaser1.pdf.

Godin, J.-G. J. "Female Mating Preference for Bold Males in the Guppy, Poecilia Reticulata." *Proceedings of the National Academy of Sciences* 93, no. 19 (September 1996): 10262–67. doi:10.1073/pnas.93.19.10262.

Goffman, Erving. *Frame Analysis: An Essay on the Organization of Experience*. Boston: Northeastern University Press, 1986.

Goodwin, Brian. "Biology Is Just a Dance." In *The Third Culture: Beyond the Scientific Revolution*, edited by John Brockman. New York: Touchstone Books, 1996.

Gould, Stephen Jay. *The Mismeasure of Man*. New York: Norton, 1996.

Gould, Steven Jay, and Richard C. Lewontin. "The Spandrels of San Marco and the Panglossian Paradigm: A Critique of the Adaptationist Programme." *Proceedings of the Royal Society of London* 205, no. 1161 (1979): 581–98.

Grammer, Karl, and Randy Thornhill. "Human (Homo Sapiens) Facial Attractiveness and Sexual Selection: The Role of Symmetry and Averageness." *J. Comp. Psychol.* 108 (1994): 233–42.

Grammer, Karl, Bernhard Fink, Anders P. Møller, and Randy Thornhill. "Darwinian Aesthetics: Sexual Selection and the Biology of Beauty." *Biological Reviews* (2003), 385–407. doi::10.1017/S1464793102006085.

Greenberg, E. Peter. "Tiny Teamwork." *Nature* 424, (July 2003). http://www.dms.ufsc.br/mip7013/arquivos/6620_Tiny%20teamwork.pdf Accessed July 13, 2014.

Grice, H.P. "Logic and Conversation." In *Readings in Language and Mind*, edited by Heimir Geirsson and Michael Losonsky, 121–33. Hoboken,: Wiley, 1996.

Haldane, J. B. S. "Population Genetics." *New Biology* 18 (1955): 34–51.

Hallé, Francis. *In Praise of Plants*. Portland: Timber Press, 2002.

Hamilton III, William J., C. Busse, and K. Smith. "Adoption of Infant Orphan Chacma Baboons." *Animal Behaviour* 30, no. 1 (1982): 29–34. doi:10.1016/S0003-3472(82)80233-9.

Hamilton, William D. "The Evolution of Altruistic Behavior." *American Naturalist* 97 (1963): 354–56.

Hamilton, W., and M. Zuk. "Heritable True Fitness and Bright Birds: A Role for Parasites?" *Science* 218, no. 4570 (October 1982): 384–87. doi:10.1126/science.7123238.

Hangarter, R. P. "Gravity, Light and Plant Form." *Plant, Cell & Environment* 20, no. 6 (2008): 796–800. http://www3.interscience.wiley.com/cgi-bin/fulltext/119148364/PDFSTART.

Hansell, Michael Henry. *Bird Nests and Construction Behaviour*. Cambridge: Cambridge University Press, 2000.

Harris, Sam. *Free Will*. New York: Free Press, 2012.

Hauser, Marc. "Origin of the Mind." *Scientific American* 301 (2009): 44–51. doi:10.1038/scientificamerican0909-44.

Hawking, Stephen W. "The Quantum Mechanics of Black Holes." *Scientific American* 236, no. 1 (1977): 34–41.

———. *The Illustrated a Brief History of Time*. New York: Bantam Books, 1996.

———. *El Universo En Una Cáscara de Nuez*. Baarcelona: Crítica/Planeta, 2002.

———. "Information Loss in Black Holes." *Physical Review* D 72, no. 8 (2005). doi:10.1103/PhysRevD.72.084013.

Heal, R. D., and A. T. Parsons. "Novel Intercellular Communication System in Escherichia Coli That Confers Antibiotic Resistance between Physically Separated Populations." *Journal of Applied Microbiology* 92 (2002): 1116–22.

Hegel, G. W. F. *Hegel's Aesthetics: Lectures on Fine Art.* Edited by T. M. Knox. Oxford: Oxford University Press, 1975.

Henderson, Mark. "'Uncovered' Prehistoric Female Figure 'Earliest Piece of Erotic Art." *The Times*, 2009. http://entertainment.timesonline.co.uk/tol/arts_and_entertainment/visual_arts/article6282102.ece.

Henrich, Joseph, and Richard McElreath. "The Evolution of Cultural Evolution." *Evolutionary Anthropology* 12 (2004): 123–35. http://www.ceacb.ucl.ac.uk/cultureclub/files/2004-2005/CC2004-10-28-Henrich_McElreath.pdf.

Henshilwood, Christopher S., Francesco D'Errico, Royden Yates, Zenobia Jacobs, Chantal Tribolo, Geoff A. T. Duller, Norbert Mercier, et al. "Emergence of Modern Human Behavior: Middle Stone Age Engravings from South Africa." *Science* 295, no. 5558 (February 2002): 1278–80. doi:10.1126/science.1067575.

Hess, Elizabeth. *Nim Chimpsky: The Chimp Who Would Be Human.* New York: Bantam Books, 2008.

Heyd, Thomas. "Aesthetics and Rock Art: An Introduction." In *Aesthetics and Rock Art*, edited by Thomas Heyd and John Clegg, 37–50. Aldershot: Ashgate Publishing, Ltd., 2005.

Hinde, Robert A. "Behavioural Habituation." In *Short-Term Changes in Neural Activity and Behaviour: A Conference Sponsored by King's College Research Centre*, edited by Gabriel Horn and Robert A. Hinde. London: Cambridge University Press, 1970.

Hoffmeyer, Jesper, and Carl Emmeche. "Code-Duality and the Semiotics of Nature, On Semiotic Modeling." *Semiotica*. Mouton de Gruyter, 1991.

Hume, David. *Of the Standard of Taste, and Other Essays*. London: Bobbs-Merrill, 1965.

Humphrey, Nicholas K. "The Illusion of Beauty." *Perception* 2, no. 4 (January 1973): 429–39.

———. "The Thick Moment." In *The Third Culture: Beyond the Scientific Revolution*, edited by John Brockman. New York: Touchstone Books, 1996.

Iacoboni, Marco, Istvan Molnar-Szakacs, Vittorio Gallese, Giovanni Buccino, John C Mazziotta, and Giacomo Rizzolatti. "Grasping the Intentions of Others with One's Own Mirror Neuron System." *PLoS Biology* 3, no. 3 (March 2005): e79. doi:10.1371/journal.pbio.0030079.

Isingrini, M. "Preimaginal Learning as a Basis of Colony-Brood Recognition in the Ant Cataglyphis Cursor." *Proceedings of the National Academy of Sciences* 82, no. 24 (December 15, 1985): 8545–47. doi:10.1073/pnas.82.24.8545.

Jablonka, Eva, and Marion J. Lamb. *Evolution in Four Dimensions: Genetic, Epigenetic, Behavioral, and Symbolic Variation in the History of Life*. Cambridge: MIT Press, 2005.

Jacob, Suma, Martha K McClintock, Bethanne Zelano, and Carole Ober. "Paternally Inherited HLA Alleles Are Associated with Women's Choice of Male Odor." *Nature Genetics* 30, no. 2 (February 22, 2002): 175–79. doi:10.1038/ng830.

Jacobs, D. K., N. Nakanishi, D. Yuan, A. Camara, S. A. Nichols, and V. Hartenstein. "Evolution of Sensory Structures in Basal Metazoa." *Integrative and Comparative Biology* 47, no. 5 (September 2007): 712–23. doi:10.1093/icb/icm094.

Jakobson, Roman. "Closing Statement: Linguistics and Poetics." In *Style in Language*, edited by Thomas Sebeok, 350–77. New York: Wiley, 1960.

———. *Essais de Linguistique Générale: Rapports Externes Du Langage*. Paris: Les Éditions de Minuit, 1973.

James, William. "What Is an Emotion?" *Mind, University Press on Behalf of the Mind Association* 9, no. 34 (1884): 188–205. Accessed July 14, 2014. http://www.jstor.org/stable/2246769.

Jang, J.-C., and J. Sheen. "Sugar Sensing in Higher Plants." *Trends Plant Sci 2*, 1997, 208–14.

Jansson, Liselotte, and Magnus Enquist. "Receiver Bias for Colourful Signals." *Animal Behaviour* 66, no. 5 (2003): 965–71. doi:DOI: 10.1006/anbe.2003.2249.

Johnsgard, Paul A. "Sympatry Changes and Hybridization Incidence in Mallards and Black Ducks." *American Midland Naturalist* 77, no. 1 (1967): 51 – 63. http://www.jstor.org/stable/2423425.

Johnson, Steven. *Sistemas Emergentes. O que tienen en comun hormigas, neuronas, ciudades y software.* Mexico: Fondo de Cultura Economica, 2003.

Kahneman, Daniel, and Amos Tversky, "On the Psychology of Prediction," *Psychological Review* 80, no. 4 (1973): 237–51. doi:10.1037/h0034747.

Kahneman, Daniel, Ed Diener, and Norbert Schwarz. *Well-Being: The Foundations of Hedonic Psychology*. New York: Russell Sage Foundation, 2003.

Kaila, Ville R. I., and Arto Annila. "Natural Selection for Least Action." *Proceedings of the Royal Society A: Mathematical, Physical and Engineering Sciences* 464, no. 2099 (2008): 3055–70. doi:10.1098/rspa.2008.0178.

Kant, Immanuel. *The Critique of Judgement*. Trans. James Creed Meredith. Electronic version, 1790. American Philosophical Association Gopher. Accessed July 13, 2014. http://eserver.org/philosophy/kant/critique-of-judgment.txt.

Kaplan, Frédéric. "Semiotic Schemata: Selection Units for Linguistic Cultural Evolution." *Proceedings of Artificial Life VII* (2000). http://www3.isrl.illinois.edu/~junwang4/langev/localcopy/pdf/kaplan00semioticSchemata.pdf.

Kaplan, Stephen. "Environmental Preferences in Knowledge-Seeking, Knowledge-Using Organims." In *The Adapted Mind*, edited by Jerome H. Barkow, Leda Cosmides, and John Tooby, 581–600. Oxford; New York: Oxford University Press, 1992.

Kauffman, Stuart A. *The Origins of Order: Self.* New York: Oxford University Press US, 1993.

Kawai, Masao. "Newly-Acquired Pre-Cultural Behavior of the Natural Troop of Japanese Monkeys on Koshima Islet." *Primates* 6, no. 1 (August 1965): 1–30. doi:10.1007/BF01794457.

Kawamura, Syunzo. "The Process of Sub-Culture Propagation among Japanese Macaques." *Primates* 2, no. 1 (March 1959): 43–60. doi:10.1007/BF01666110.

King, John. *Reaching for the Sun: How Plants Work*. Cambridge: Cambridge University Press, 1997.

Knight, Chris. "Darwinism and Collective Representations." In *The Archaeology of Human Ancestry: Power, Sex, and Tradition*, edited by James Steele and Stephen Shennan, 5:331–60. New York: Routledge, 1996. doi:http://www.radicalanthropologygroup.org/pub_knight_power_watts.pdf.

———. *Ritual Speech Coevolution; in a Solution to the Problem of Deception Approaches to the Evolution of Language: Social and Cognitive Bases*. Edited by Chris Hurford, James R. Studdert-Kennedy, Michael Knight. Cambridge: Cambridge University Press, 1998.

Knight, Chris, Camilla Power, and Ian Watts. "The Human Symbolic Revolution: A Darwinian Account." *Cambridge Archaeological Journal* 5, no. 01 (1995): 75–114. doi:10.1017/S0959774300001190.

Koestler, Arthur. *The Act of Creation*. London: Hutchinson, 1976.

Komar, Vitaly, and Aleksandr Melamid. *Painting by Numbers: Komar and Melamid's Scientific Guide to Art*, edited by JoAnn Wypijewski. Berkeley: University of California Press, 1999.

Korzeniewski, B. "Cybernetic Formulation of the Definition of Life." *Journal of Theoretical Biology* 209, no. 3 (2001): 275–86. doi:10.1006/jtbi.2001.2262.

Krampen, Martin. "Phytosemiotics." *Semiotica—Journal of the International Association for Semiotic Studies / Revue de l'Association Internationale de Sémiotique* 36, no. 3–4 (1981): 187–210. doi:doi:10.1515/semi.1981.36.3-4.187.

Krebs, John R. "Territory and Breeding Density in the Great Tit, Parus Major L." *Ecology* 52, no. 1 (1971): 3 – 22.

Krebs, John R., and Richard Dawkins. "Animal Signals: Mind Reading and Manipulation." In *Behavioural Ecology: An Evolutionary Approach*, edited by John R. Krebs and Nicholas B. Davies. Malden, MA: Blackwell Publishing, 1984.

———. "Animal Signals: Information or Manipulation." Edited by J. R. Krebs And N. B. Davies. *Behavioural Ecology: An Evolutionary Approach*. Malden, MA: Blackwell Publishing, 1978.

Kroeber, A. L. "Stimulus Diffusion." *American Anthropologist, New Series* 42, no. 1 (1940).

———. "The Superorganic." *American Anthropologist, New Series* 19, no. 2 (1917): 163–213. http://www.jstor.org/stable/660754.

Kull, Kalevi. "Evolution and Semiotics." In *The Semiotic Web*, edited by Thomas Albert Sebeok and Donna Jean Umiker-Sebeok, 221–34. New York: Mouton de Gruyter, 1992.

———. "Semiotic Ecology: Different Natures in the Semiosphere." *Sign Systems Studies* 1, no. 1 (1998): 344–71. http://www.ceeol.com/aspx/issuedetails.aspx?issueid=6f1f6605-909d-415d-b862-acbf968ba2c5&articleId=1a9d3cc9-8800-4213-896c-653f52ce1b23.

———. "On the History of Joining Bio with Semio: F. S. Rothschild and the Biosemiotic Rules." *Sign Systems Studies*, no. 1/1999 (1999): 128–38.

Kurland, Philip B., and Ralph Lerner, eds. *The Founders' Constitution*. Chicago: University of Chicago Press, 1987. Accessed February 28, 2010. http://press-pubs.uchicago.edu/founders/

Lakoff, George, and Mark Johnson. *Metaphors We Live by*. Chicago: University of Chicago Press, 1980.

Laland, Kevin N., and Gillian R. Brown. *Sense and Nonsense: Evolutionary Perspectives on Human Behaviour*. Oxford; New York: Oxford University Press, 2002.

LeVay, S. "A Difference in Hypothalamic Structure between Heterosexual and Homosexual Men." *Science* 253, no. 5023 (August 1991): 1034–37. doi:10.1126/science.1887219.

Levitin, Daniel J., and Vinod Menon. "The Neural Locus of Temporal Structure andExpectancies in Music: Evidence From Functional Neuroimaging At 3 Tesla ." *Music Perception,* 22, no. 3 (2005): 563–75. doi:doi:10.1525/mp.2005.22.3.563.

Lewontin, Richard C. *The Triple Helix: Gene, Organism, and Environment*. Cambridge: Harvard University Press, 2001.

Lewontin, Richard C., Steven Rose, and Leon J. Kamin. *Not in Our Genes: Biology, Ideology, and Human Nature*. New York: Pantheon Books, 1985.

Lloyd, Seth, and Y. Jack Ng. "Black Hole Computers." *Scientific American*, 2004, 31–39.

Logothetis, N. K. "Single Units and Conscious Vision." *Philosophical Tranasactions of the Royal Society of London* 353 (1998): 1801–18.

Lotman, Juri. "On the Semiosphere." *Sign Systems Studies* 33, no. 1 (2005): 5–23. http://www.ut.ee/SOSE/sss/Lotman331.pdf.

Madden, J. "Preferences for Coloured Bower Decorations Can Be Explained in a Nonsexual Context." *Animal Behaviour* 65, no. 6 (June 2003): 1077–83.

Malinowski, Bronislaw. *Argonauts of the Western Pacific*. New York: Read Books, 2007.

Mandel, David. *Changing Art, Changing Man.* New York: Horizon Press, 1967.

Mandoki, Katya. *Prosaica: Introducción a La Estética de Lo Cotidiano*. Mexico: Grijalbo, 1994.

———. "Sites of Symbolic Density: A Relativistic Approach to Experienced Space." In *Philosophies of Place III*, edited by Andrew Light. New York: Rowman & Littlefield, 1998.

———. "A Host of Ghosts in Descartes's Theater." *Semiotica*, 2002.

———. "Semiosis Y Aisthesis: ¿cinta Möbius O Doble Hélice?" In *Memorias Del VII Congreso Internacional de La Asociación Internacional de Semiótica Visual. CD Rom.*, edited by AMESVE, 2004.

———. *Estética Cotidiana Y Juegos de La Cultura: Prosaica 1*. México, DF: Siglo XXI Editores, 2006.

———. *Everyday Aesthetics: Prosaics, the Play of Culture and Social Identities*. Aldershot UK: Ashgate Pub Co, 2007.

———. "Culturomes and Negentropic Processes: A Biosemiotic Approach." In *Communication Understanding. International Semiotics Institute, IASS-AIS World Congress of Semiotics*, edited by Eero Tarasti (Ed.), 1005–12. Helsinki Imatra: Semiotic Society of Finland Helsinki, 2009.

———. "Zoo-pragmatics: Performative Acts among Animals." In *Linguistic Aspects of Intercultural Pragmatics*, edited by Istvan Kecskes and Jesus Romero-Trillo, 421–38. Berlin and New York: Mouton de Gruyter, 2013.

Mandoki Winkler, Catalina Inés. *Estética Y Poder.* Universidad Nacional Autónoma de México, 1991.

Maran, Timo. "John Maynard Smith's Typology of Animal Signals: A View from Semiotics." *Sign Systems Studies* 37, no. 33–34 (2009).

Margulis, Lynn. "Gaia Is a Tough Bitch." In *Third Culture: Beyond the Scientific Revoluti…*, edited by John Brockman. First Touchstone Edition, 1996.

———. "Genetic and Evolutionary Consequences of Symbiosis." *Experimental Parasitology* 39, no. 2 (April 1976): 277–349. doi:10.1016/0014-4894(76)90127-2.

Martindale, Colin, and David Rindos. "On Hedonic Selection, Random Variation, and the Direction of Cultural Evolution." *Current Anthropology* 27, no. 1 (1986): 50–53.

Martinelli, Dario. "Liars, Players, and Artists: A Zoösemiotic Approach to Aesthetics." *Semiotica— Journal of the International Association for Semiotic Studies / Revue de l'Association Internationale de Sémiotique* 150, no. 1/4 (2004.): 77–118.

Marx, Karl. "The Fetishism of Commodities and the Secret Thereof." In *Capital: A Critique of Political Economy*, 81–96. New York: The Modern Library, 1906.

Matsuzawa, T, and G. Yamakoshi. "Comparison of Chimpanzee Material Culture Between Bossou and Nimba, West Africa." In *Reaching into Thought: The Minds of the Great Apes*, edited by Sue Taylor Parker, Anne E. Russon, Kim A. Bard. Cambridge University Press, 1996.

Maturana, Humberto R., and Francisco J. Varela. *The Tree of Knowledge: The Biological Roots of Human Understanding*. Boston: Shambhala Publications, 1992.

Mauss, Marcel. *The Gift: The Form and Reason for Exchange in Archaic Societies*. New York: Routledge, 1990.

Merrell, Floyd. *Signs Grow: Semiosis and Life Processes*. Toronto: University of Toronto Press, 1996.

Miller, Geoffrey F. *The Mating Mind: How Sexual Choice Shaped the Evolution of Human Nature*. New York: Anchor Books, 2001.

Miller, Patrick, Nicoletta Biassoni, Amy Samuels, and Peter L. Tyack. "Whale Songs Lengthen in Response to Sonar." *Nature* 405 (2000). http://foodweb.uhh.hawaii.edu/MARE390_files/Milleretal.2000.pdf.

Mithen, Steven J. *The Prehistory of the Mind: The Cognitive Origins of Art, Religion and Science*. London: Thames and Hudson, 1999.

———. *The Singing Neanderthals: The Origins of Music, Language, Mind, and Body*. Cambridge: Harvard University Press, 2005.

Moller, A.P., and A. Pomiankowski. "Why Have Birds Got Multiple Sexual Ornaments?" *Behavioral Ecology and Sociobiology* 32, no. 3 (March 1993): 167–76. doi:10.1007/BF00173774.

Montagu, Ashley. *Touching: The Human Significance of the Skin*. New York: Harper & Row, 1978.

Morel, Laurence, Robert K. Meer, and Barry K. Lavine. "Ontogeny of Nestmate Recognition Cues in the Red Carpenter Ant (Camponotus Floridanus)." *Behavioral Ecology and Sociobiology* 22, no. 3 (March 01, 1988): 175–83. doi:10.1007/BF00300567.

Moreno, Juan, Manuel Soler, Anders Pape Møller, and Mats Linden. "The Function of Stonecarrying in the Black Wheatear, Oenanthe Leucura." *Animal Behaviour* 47, no. 6 (1994).

Morris, Desmond. *The Naked Ape; a Zoologist's Study of the Human Animal*. New York: McGraw-Hill, 1967.

Mott, K. A. "Sensing of Atmospheric CO2 by Plants." *Plant, Cell, and Environment* 13(1990): 731–37.

Mukerjee, Madhusree. "Circles for Space:" *Scientific American* 289, no. 6 (2003): 32–34.

Nadal, Marcos, Enric Munar, Jaume 1 Capó, Miquel Àngel Rosselló, and Camilo José Cela-Conde. "Towards a Framework for the Study of the Neural Correlates of Aesthetic Preference." *Spatial Vision* 21 no. 3–5 (2008): 379–96. http://evocog.org/images/stories/documentos/publicaciones/2008_neural_correlates.pdf.

Nahon, E., D. Atzmony, A. Zahavi, and D. Granot. "Mate Selection In Yeast: A Reconsideration of the Signals and the Messages Encoded by Them." *Journal of Theoretical Biology* 172, no. 4 (1995): 315–22.

Needham, Joseph, Kenneth Girdwood Robinson, and Ray Huang. *Science and Civilisation in China: General Conclusions and Reflections*. New York: Cambridge University Press, 2004.

Norman, Mark D., Julian Finn, and Tom Tregenza. "Dynamic Mimicry in an Indo-Malayan Octopus." *Proceedings of the Royal Society B* 268, no. 1478 (2001): 1755–58.

Norman, Mark D., and F. G. Hochberg. "The 'Mimic Octopus' (Thaumoctopus Mimicus N. Gen. et Sp.), a New Octopus from the Tropical Indo-West Pacific (Cephalopoda: Octopodidae)." *Molluscan Research* 25, no. 2 (2005): 57–70.

Olds, J., and P. Milner. "Positive Reinforcement Produced by Electrical Stimulation of Septal Area and Other Regions of Rat Brain." *Psychological Review* 47 (1954): 419–27.

Orians, Gordon H., and Judith H. Heerwagen. "Evolved Response to Landscapes." In *The Adapted Mind*, edited by Jerome H. Barkow, Leda Cosmides, and John Tooby. Oxford, New York: Oxford University Press, 1992.

Ostlund-Nilsson, S., and Mikael Holmlund. "The Artistic Three-Spined Stickleback (*Gasterosteous aculeatus*)." *Behavioral Ecology and Sociobiology* 53, no. 4 (2003): 214–20. doi:10.1007/s00265-002-0574-z.

Packard, Vance. *The Hidden Persuaders*. New York: Penguin Books, 1991.

Parret, Herman. *De La Semiótica a La Estética: Enunciación, Sensación, Pasiones*. Edited by lingüística, comunicación Colección Edicial Universidad. Lengua. Buenos Aires: Edicial, 1995.

———. *The Aesthetics of Communication: Pragmatics and beyond*. Aphen aan der Rijn: Kluwer Academic Publishers, 1993.

Patel, Aniruddh D., John R. Iversen, Micah R. Bregman, Irena Schulz, and Charles Schulz. "Investigating the Human-Specificity of Synchronization to Music." M. Adachi et al. (Eds.), *Proceedings of the 10th Intl. Conf. on Music Perception and Cognition, August 2008, Sapporo, Japan.* Adelaide: Causal Productions, 2008. http://birdloversonly.org/docs/Patel.pdf.

Pavlov, I. P. *Lectures on Conditioned Reflexes*. London: Allen and Unwin, 1928.

Payne, R. B., L. L. Payne, J. L. Woods, and M. D. Sorenson. "Imprinting and the Origin of Parasite-Host Species Associations in Brood-Parasitic Indigobirds, Vidua Chalybeata." *Animal Behaviour* 59, no. 1 (2000): 69–81. doi:10.1006/anbe.1999.1283.

Peirce, Charles Sanders. *Philosophical Writings of Peirce*. Edited by Justus Buchler. Courier London: Dover Publications, 1955.

Pert, Candace B. *Molecules of Emotion: Why You Feel the Way You Feel*. New York: Simon & Schuster, 1999.

Philip, Chase, and Harold L Dibble. "Middle Paleolithic Symbolism: A Review of Current Evidence and Interpretations." *Journal of Anthropological Archaeology* 6, no. 3 (September 1987): 263–96. doi:10.1016/0278-4165(87)90003-1.

Piaget, Jean. *Play, Dreams and Imitation in Childhood*. New York: Routledge, 1999.

Pinker, Steven. *The Language Instinct*. New York: W. Morrow and Co., 1994.

———. *How the Mind Works*. New York: Norton, 1997.

Pollack, Robert. *Signs of Life: The Language and Meanings of DNA*. Boston: Houghton Mifflin, 1994.

Portmann, Adolf. *Animal Forms and Patterns*. New York: Schocken Books, 1967.

Powell, Camilla, and Leslie Aiello. "Female Proto-Symbolic Strategies." In *Women in Human Evolution*, edited by Lori D. Hager, 153. New York: Routledge, 1997.

Prather, J. F., S. Peters, S. Nowicki, and R. Mooney. "Precise Auditory-Vocal Mirroring in Neurons for Learned Vocal Communication." *Nature* 451, no. 7176 (January 17, 2008): 305–10. doi:10.1038/nature06492.

Prosser, C. Ladd. "Modes of Communication." In *Comparative Neurobiology*, edited by Melvin J. Cohen and Felix Sturmwasser. New York: John Wiley, 1985.

Ramachandran, S., and E. M. Hubbard. "I: Robustness of Synaesthetic Effects." *Journal of Consciousness Studies* 10, no. 8 (2003): 49–57.

Ramakrishnan, U., R. G. Coss, J. Schank, A. Dharawat, and S. Kim. "Snake Species Discrimination by Wild Bonnet Macaques (Macaca Radiata)." *Ethology* 111 (2005): 337–56.

Reader, S. M., and K. N. Laland. "Do Animals Have Memes?" *Journal of Memetics—Evolutionary Models of Information Transmission* 3 (1999).

Rees, Martin. "An Ensemble of Universes." In *The Third Culture: Beyond the Scientific Revolution*, edited by John Brockman, 413. New York: Touchstone Books, 1996.

Reid, Thomas. *Essays on the Active Powers of Man*. Printed for J. Bell, 1801. accessed July 13, 2014. http://books.google.co.il/books?id=kCA-AAAAcAAJ&printsec=frontcover&source=gbs_ge_summary_r&cad=0#v=onepage&q&f=false.

Ricardo, Alonso, and Jack W. Szostak. "Origin of Life on Earth." *Scientific American* (September 2009): 38.

Richerson, Peter J., and Robert Boyd. *Not by Genes Alone: How Culture Transformed Human Evolution*. Chicago: University of Chicago Press, 2005.

Ritchie, Michael G. "The Inheritance of Female Preference Functions in a Mate Recognition System." *The Royal Society* 267 (2000): 327–32. Accessed July 13, 2014. http://biology.st-and.ac.uk/supplemental/ritchie/papers/InheritanceMRS.pdf.

Rizzolatti, G., and L. Craighero. "The Mirror-Neuron System." *Annual Review of Neuroscience* 27 (2004): 169–92.

Rohwer, Sievert. "Status Signaling in Harris Sparrows: Some Experiments in Deception." *Behaviour* 61, no. 1 (1977): 107–29.

Rossi-Landi, Ferruccio. *Language as Work & Trade: A Semiotic Homology for Linguistics & Economics*. Westport: Bergin & Garvey Publishers, 1983.

Rothenberg, David. *Why Birds Sing: A Journey Into the Mystery of Bird Song*. New York: Basic Books, 2006.

Rothschild, Friedrich Salomon. *Creation and Evolution: a Biosemiotic Approach*. New Jersey: Transaction Publishers, 1994.

Ryan, Michael, J. "Female Mate Choice in a Neotropical Frog." *Science* 25, vol. 209, no. 4455 (July 1980): 523–25. doi: 10.1126/science.209.4455.523

———. "Sexual Selection, Receiver Biases, and the Evolution of Sex Differences." *Science* 281, no. 5385 (September 1998): 1999–2003. doi:10.1126/science.281.5385.1999.

Sachs, Curt. *World History of Dance*. New York: Norton, 1937.

Sachs, T. "Patterned Differentiation in Plants." *Differentiation* 11 (1978): 65–73.

Sagan, L. "On the Origin of Mitosing Cells." *Journal of Theoretical Biology* 14, no. 3 (March 1967): 225–IN6. doi:10.1016/0022-5193(67)90079-3.

Sahlins, Marshall. *The Use and Abuse of Biology*. Ann Arbor: University of Michigan Press, 1976. http://www.raco.cat/index.php/Papers/article/viewFile/24857/58316.

Salthe, Stanley N. "What Is the Scope of Biosemiotics? Information in Living Systems." In *Introduction to Biosemiotics; The New Biological Synthesis*, 133–48. New York: Springer, 2008.

Sánchez Vázquez, Adolfo. *Invitación a La Estética*. Mexico: Grijalbo, 1992.

Schank, Roger. "Information Is Surprises." In *The Third Culture: Beyond the Scientific Revolution*, edited by John Brockman, 168–76. New York: Touchstone Books, 1996.

Sebeok, Thomas Albert. "Review of Communication among Social Bees; Porpoises and Sonar; Man and Dolphin." *Language* 39, no. 3 (1963): 448–66.

———. "The Semiotic Web: A Chronicle of Prejudices." *Bulletin of Literary Semiotics* 34 (1975).

———. *Perfussion of Signs*. Bloomington: Indiana University Press, 1977.

———. *Sight, Sound, and Sense*. Bloomington: Indiana University Press, 1978.

———. "Prefigurements of Art." *Semiotica—Journal of the International Association for Semiotic Studies / Revue de l'Association Internationale de Sémiotique* 27, no. 1–3 (1979): 3–72.

———. "Naming in Animals, with Reference to Playing: A Hypothesis." In *Essays in Zoosemiotics*, 77–92. Toronto: Victoria College in the University of Toronto, 1990.

———. "Semiotics and the Biological Sciences." *Indiana University*, 1995.

———. *Signos: Una Introducción a La Semiótica*. Barcelona: Paidos, 1996.

Accessed July 14 2014. http://www.scribd.com/doc/43683005/Sebeok-Semiotics-and-the-Biological-Sciences.

———. "The Music of the Spheres." *Semiotica* 128, no. 3–4 (2000): 527–36.

———. "The Sign Science and the Life Science." *Applied Semiotics / Sémiotique Appliquée* 3, no. 6/7 (1999): 85–96. Accessed July 14 2014. http://french.chass.utoronto.ca/as-sa/ASSA-6-7/Vol3.No6-7.Sebeok.pdf.

———. *Global Semiotics*. Bloomington: Indiana University Press, 2001.

Sebeok, Thomas Albert, and Diana Agrest. *A Perfusion of Signs*. Vol. 1975. Bloomington: Indiana University Press, 1975.

Sebeok, Thomas Albert, and Donna Jean Umiker-Sebeok. *The Semiotic Sphere*. New York: Plenum Press, 1986.

Sebeok, Thomas Albert, Donna Jean Umiker-Sebeok, and Evan P. Young. *Biosemiotics: The Semiotic Web 1991*. Edited by Thomas Albert Sebeok, Donna Jean Umiker-Sebeok, and Evan P. Young. The Hague: Walter de Gruyter, 1992.

Sebeok, Thomas Albert, and Jean Umiker-Sebeok. *The Semiotic Web 1986*. Edited by Thomas Albert Sebeok and Jean Umiker-Sebeok. The Hague:Walter de Gruyter, 1987.

Sharov, Alexei A. "Information, Biosemiotics: Functional-Evolutionary Approach to the Analysis of the Sense." In *Biosemiotics. The Semiotic Web 1991*, edited by Thomas Sebeok and Jane Umiker-Sebeok, 345–73. The Hague:Walter de Gruyter, 1992.

Shaw. "Feeling of Beauty Among Animals." *Athenæum* (1866): 681. Accessed July 6 2013. http://darwin-online.org.uk/content/frameset?viewtype=image&itemID=CUL-DAR84.1.181&pageseq=1.

Sheen, J., L. Zhou, and J.-C. Jang. "Sugars as Signalling Molecules." *Current Opinions in Plant Biology* 2, no. 410–18 (1999).

Sheets-Johnstone, Maxine. *The Roots of Power: Animate Form and Gendered Bodies*. New York: Open Court Publishing, 1994.

Sherman, Paul W., and Hudson Kern Reeve. "Sexual Selection and Sensory Exploitation." *Science* 283, no. 5405 (1999): 1083a–1083. doi:10.1126/science.283.5405.1083a.

Shusterman, Richard. *Body Consciousness: A Philosophy of Mindfulness and Somaesthetics*. New York: Cambridge University Press, 2008.

Simmel, Georg. *Simmel on Culture: Selected Writings*. Edited by David Frisby and Mike Featherstone. Thousand Oaks: Sage Publications, 1997.

Singer, M. C., D. Ng, and C. D. Thomas. "Heritability of Oviposition Preference and Its Relationship to Offspring Performance within a Single Insect Population." *Evolution* 42, no. 5 (1988): 977–85.

Smith, Adam. *The Wealth of Nations. Book V Ch.ii Part II, Article IV. Taxes upon Consumable Commodities*. Edited by Edwin Canaan. Chicago: Chicago University Press, 1776. Accessed July 13, 2014. http://www.staff.ncl.ac.uk/j.veit-wilson/documents/adamsmithontaxesandnecessities.pdf

Smith, Fabienne. "Charles Darwin's Ill Health." *Journal of the History of Biology* 23, no. 3 (1990): 443–59. doi:10.1007/BF00136378.

Smith, John Maynard, and David Harper. "Animal Signals: Models and Terminology." *Journal of Theoretical Biology* 177, no. 3 (1995): 305–11.

Smith, John Maynard, and Eörs Szathmáry. *The Major Transitions in Evolution*. Oxford; New York: Oxford University Press, 1997

Smolin, Lee. "A Theory of the Whole Universe." In *The Third Culture: Beyond the Scientific Revolution*, edited by John Brockman. New York: Touchstone Books, 1996.

Snowdon, Charles T, and David Teie. "Affective Responses in Tamarins Elicited by Species-Specific Music." *Biology Letters* 6, no. 1 (2010): 30–32. doi:10.1098/rsbl.2009.0593.

Sonea, Sorin. "Half of the Living World Was Unable to Communicate for About a Billion Years." *Semiotic Web 1991*. Berlin: Walter de Gruyter, 1992.

Sperber, Dan. *Explaining Culture: A Naturalistic Approach*. Oxford and Massachusetts: Blackwell, 1996.

Stenger, Victor J. *The Fallacy of Fine-Tuning: Why the Universe Is Not Designed for Us*. Amherst New York: Prometheus Books, 2011.

———. "The Anthropic Principle." *The Encyclopedia of Nonbelief*, n.d.

Sütterlin, Christa. "Art, from Sign and Schema to Iconic Representation: Evolutionary Aesthetics of Pictorial." In *Evolutionary Aesthetics*, edited by Eckart Voland and Karl Grammer, 377. New York: Springer, 2003.

Swenson, Rod. "Epistemic Ordering and the Development of Space-Time: Intentionality as a Universal Entailment." *Semiotica* 127, no. 1/4 (1999): 567–97.

Symons, Donald. *The Evolution of Human Sexuality*. Oxford; New York: Oxford University Press, 1979.

Szostak, Jack W. "Evolution Ex Vivo." *Nature* (1993). doi:10.1038/361119a0.

Szostak, Jack W., David P. Bartel, and P. Luigi Luisi. "Synthesizing Life." *Nature* 409, no. 18 (2001). Accessed July 14 2014. https://www.biophysics.org/Portals/1/PDFs/Summer%20Course/Synthesizing%20Life%20-%20Lecture%2015.pdf.

Tallis, Raymond. "The Limitations of a Neurological Approach to Art." *The Lancet* 372, no. 9632 (2008): 19–20. doi:10.1016/S0140-6736(08)60975-7.

Terkel, Joseph. "Cultural Transmission of Feeding Behavior in the Black Rat (Rattus Rattus). Social Learning in Animals: The Roots of Culture." In *Social Learning in Animals: The Roots of Culture*, 17–47. San Diego: Academic Press, 1996.

Thompson, D'Arcy W. *On Growth and Form*. New York: Macmillan, 1946. Accessed July 14 2014. http://ia600301.us.archive.org/10/items/ongrowthform00thom/ongrowthform00 thom.pdf.

Thornhill, Randy. "Darwinian Aesthetics Informs Traditional Aesthetics." In *Evolutionary Aesthetics*, edited by Eckart Voland and Karl Grammer, 9–35. New York: Springer, 2003.

Tinbergen, Nico. "On the Analysis of Social Organization Among Vertebrates, with Special Reference to Birds." *American Midland Naturalist* 21, no. 1 (1939): 210–34.

Tooby, John, and Leda Cosmides. "Psychological Foundations of Culture." In *The Adapted Mind*, edited by Jerome H. Barkow, Leda Cosmides, and John Tooby. Oxford; New York: Oxford University Press US, 1992.

Trivers, Robert. "The Evolution of Reciprocal Altruism." *The Quarterly Review of Biology* 46, no. 1 (1971): 35 – 57.

———. *Social Evolution*. Menlo Park: Benjamin/Cummings Pub. Co., 1985.

Turner, Mark. "The Art of Compression." In *The Artful Mind: Cognitive Science and the Riddle of Human Creativity*, edited by Mark Turner, 94–113. Oxford; New York: Oxford University Press, 2006.

Uršič, Marko. "Cogito Ergo Mundus Talis Est." *Acta Analytica* 17, no. 1 (March 1, 2002): 53–67. doi:10.1007/BF03177507.

Varela, Francisco J. "The Emergent Self." In *The Third Culture: Beyond the Scientific Revolution*, edited by John Brockman, 210–16. New York: Touchstone Books, 1996.

Varela, Francisco J., Evan Thompson, and Eleanor Rosch. *The Embodied Mind: Cognitive Science and Human Experience*. Cambridge: MIT Press, 1992.

Veblen, Thorstein. *The Theory of the Leisure Class*. New York: Forgotten Books, 1924.

Vélez Caicedo, Ana Cristina. *Homo Artisticus: Una Perspectiva Biológica-Evolutiva*. 1st ed. Medellín: Editorial Universidad de Antioquia, 2008.

Vernadskiĭ, Vladimir Ivanovich, and Mark A. McMenamin. *The Biosphere*. New York: Springer, 1998.

Voland, Eckart, and Karl Grammer. *Evolutionary Aesthetics*. Edited by Eckart Voland and Karl Grammer. Dordrecht: Springer, 2003.

Voltaire, Francois. *Philosophical Dictionary*. Edited by Theodore Besterman. New York: Penguin Group 1984.

von Frisch, Karl. *The Dancing Bees: An Account of the Life and Senses of the Honeybee*. London: Methuen and Company Ltd., 1954.

———. *Bees: Their Vision, Chemical Senses, and Language*. Ithaca: Cornell University Press, 1950.

von Uexküll, Jakob. "The Theory of Meaning. ." *Semiotica* 42, no. 1 (1982): 25–82.

———. "A Stroll through the Worlds of Animals and Men: A Picture Book of Invisible Worlds (1934)." *Semiotica* 89., no. 4 (1992): 319–91.

von Uexküll, Thure, Geigges Werner, and Jörg M. Herrmann. "Endosemiosis." *Semiotica* 96, no. 1–2, (1993): 5–52. doi:10.1515/semi.1993.96.1-2.5//1993.

Wallace, Alfred Russel. "Darwinism An Exposition of The Theory of Natural Selection With Some of Its Applications." *Project Gutenberg*, 1889.

———. "Review of Origin 4th Edn. Mimicry, and Other Protective Resemblances among Animals." *Westminster and Foreign Quarterly Review* 88–173, no. 32–1 (1867): 1–43.

Washi, Rumi. "'Japanese Female Speech' and Language Policy in World War II Era." In *Japanese Language, Gender, and Ideology: Cultural Models and Real People*, edited by Shigeko Okamoto and Janet S. Shibamoto Smith. Oxford; New York: Oxford University Press, 2004.

Weiss, Allen S. *The Aesthetics of Excess*. New York: SUNY Press, 1989.

Weitz, Morris. "The Role of Theory in Aesthetics." In *Philosophy Looks at the Arts*, edited by Joseph Margolis, 3rd ed. Philadelphia: Temple University Press, 1987.

Welsch, Wolfgang. "Animal Aesthetics." *Contemporary Aesthetics* 2 (2004). http://www.contempaesthetics.org/newvolume/pages/article.php?articleID=243).

Whitehead, N. *An Enquiry Concerning the Principles of Natural Knowledge*. Cambridge, New York: Cambridge University Press, 1919.

Whiten, A., J. Goodall., W. C. McGrew, T. Nishida, V. Reynolds, Y. Sugiyama, C. E. G. Tutin, R. Wrangham, and W. C. Boesch. "Cultures in Chimpanzees." *Nature* 399 (June 1999): 682–85.

Wickler, Wolfgang. "Mimicry in Tropical Fishes." *Philosophical Transactions of the Royal Society of London. Series B, Biological Sciences* 251, no. 772 (1966): 473–74.

Wiley, Norbert. *The Semiotic Self*. Chicago: University of Chicago Press, 1994.

Williams, George C. *Adaptation and Natural Selection.* Princeton, New Jersey: Princeton University Press , 1966.

Wilson, Edward O. *Sociobiology*. Cambridge: Harvard University Press, 1980.

Wreschner, Ernst E. "Red Ochre and Human Evolution: A Case for Discussion." *Current Anthropology* 21, no. 5 (1980): 631.

Wright, Robert. *The Evolution of God*. New York: Little, Brown & Co., 2009.

———. *The Moral Animal: Evolutionary Psychology and Everyday Life*. New York: Pantheon Books, 1994.

Zahavi, Amotz. "Mate selection: A Selection for a Handicap." *Journal of Theoretical Biology* 53, no. 1 (September 1975): 205–14. doi:10.1016/0022-5193(75)90111-3.

Zahavi, Amotz, and Avishag Zahavi. *The Handicap Principle: A Missing Piece of Darwin's Puzzle*. Oxford: Oxford University Press, 1997.

Zeki, Semir. "The Representation of Colours in the Visual Cortex." *Nature* 284 (1980): 412–18.

———. "The Neurology of Ambiguity." In *The Artful Mind: Cognitive Science and the Riddle of Human Creativity*, edited by Mark Turner, 243–70. Oxford: Oxford University Press, 2006.

Zeki, Semir, and John Paul Romaya. "Neural Correlates of Hate." *PloS One* 3, vol. 10 (2008): e3556. doi:10.1371/journal.pone.0003556.

Index

About the Author

Katya Mandoki is a scholar of philosophy, an author, and an experimental artist who initiated in the early nineties the systematic research of everyday aesthetics or *prosaics*, publishing seven books in this field of inquiry: *Prosaica: Introducción a la Estética de lo Cotidiano* (1994), *Everyday Aesthetics: Prosaics, the Play of Culture, and Social Identities* (2007), *El Indispensable Exceso de la Estética* (2013), *Estética Cotidiana y Juegos de la Cultura* (2006), *Prácticas Estéticas e Identidades Sociales* (2006), and *Estética y Comunicación: de Acción, Pasión y Seducción* (2006). Mandoki was nationally awarded on two occasions for her artworks, one of which is Histogram a monumental installation permanently exhibited at the Library Plaza in Campus Xochimilco of the Metropolitan Autonomous University in Mexico. She has been professor of aesthetics and semiotics at this institution, a member of the Mexican Academy of Sciences and of the National System of Researchers, and founder and Honorary Member of the Mexican Association of Aesthetics.

www.ingramcontent.com/pod-product-compliance
Lightning Source LLC
LaVergne TN
LVHW101446200926
848483LV00001B/63
* 9 7 8 1 4 9 8 5 0 3 0 6 8 *